AF797730

PRAISE FOR TERENCE WRIGHT

From *"The Power of Hallucinogens:"*

The author provides very detail information on both the physical and historical aspects of each hallucinogen. The book gives you the risks and the benefits of each substance as well. I have heard about all of the hallucinogens the author talks about, but have had not personal use of any of them. I have often been curious about a few of these and feel that the author provided some really great information. I especially like the personal stories towards the end. I highly recommend this book to anyone who has an interest in hallucinogens and their properties.

— MISSY RICHARDSON

This book provides a comprehensive and informative overview of the history and use of psychedelics. The author does a great job of presenting both the benefits and the potential risks of these substances in a balanced and objective manner.

— STOICPRENEUR

"The Power of Hallucinogens" delves deep into the concept of using hallucinogenics in a safe and responsible manner. This book details all of the main types of hallucinogenics, going into their histories, design, intended use, chemical composition, and how-to safely use them. Definitely recommended!

— KYLE WALLER

THE POWER OF DISSOCIATIVES

THE POWER OF DISSOCIATIVES

LEARN ABOUT KETAMINE, PCP, DXM, NITROUS OXIDE, MXE, 3-MEO-PCP, AND SALVIA DIVINORUM AND THEIR IMPACT ON MENTAL HEALTH, SOCIETY, AND ADDICTION

JOURNEY INTO THE PSYCHEDELIC MIND

TERENCE WRIGHT

Book
Bound Studios

Copyright © 2023 by Terence Wright

All rights reserved. No part of this book may be reproduced, stored in a retrieval system, or transmitted in any form or by any means, electronic, mechanical, photocopying, recording, or otherwise, without the prior written permission of the publisher, Book Bound Studios.

The information contained in this book is based on the author's personal experiences and research. While every effort has been made to ensure the accuracy of the information presented, the author and publisher cannot be held responsible for any errors or omissions.

This book is intended for general informational purposes only and is not a substitute for professional medical, legal, or financial advice. If you have specific questions about any medical, legal, or financial matters, you should consult with a qualified healthcare professional, attorney, or financial advisor.

Book Bound Studios is not affiliated with any product or vendor mentioned in this book. The views expressed in this book are those of the author and do not necessarily reflect the views of Book Bound Studios.

To my family and friends, who have always supported me in my endeavors.
To the indigenous cultures that have long recognized the power of these substances and to the early researchers who dared to study them.
And to the countless individuals who have had profound and transformative experiences through the use of these substances, this book is dedicated.

The only way to make sense out of change is to plunge into it, move with it, and join the dance.

— ALAN WATTS

CONTENTS

INTRODUCTION

Dissociative drugs, also known as **hallucinogens**, are a group of mind-altering substances that produce changes in perception, consciousness, and behavior. They have been used for centuries for religious and spiritual purposes, but in recent times, they have become popular recreationally. Dissociative drugs have both positive and negative effects on individuals and society, making it crucial to understand their impact.

This book provides a comprehensive overview of common dissociative drugs, including ketamine, PCP, DXM, nitrous oxide, MXE, 3-MeO-PCP, and salvia divinorum. You will learn about their historical background, chemical properties, medical uses and abuses, risks and side effects, and laws and regulations. Personal experiences with dissociative drugs are also presented to give readers a deeper understanding of the effects these substances can have. Furthermore, the book explores the relationship between dissociative drugs and mental health, society, and addiction, providing a well-rounded examination of the topic.

The target audience for this book includes anyone interested in learning about dissociative drugs, healthcare professionals, law enforcement, policymakers, and those seeking treatment for dissociative drug addiction. This book aims to educate readers about the effects of dissociative drugs and provide information that individuals can use to make informed decisions about their use. But first, let's tackle some FAQs to set the context.

What are dissociative drugs?

Dissociative drugs are a class of drugs that produce a feeling of detachment from one's environment and self. They can cause hallucinations, changes in perception, and a sense of detachment from reality.

What are the effects of dissociative drugs?

The effects of dissociative drugs vary depending on the type of drug and the dose taken. Common effects include altered perception of time and space, visual and auditory hallucinations, out-of-body experiences, and feelings of detachment from reality.

What are the risks associated with dissociative drugs?

Dissociative drugs can cause serious physical and psychological health risks, including increased heart rate, blood pressure, body temperature, nausea, vomiting, confusion, and paranoia. In addition, long-term use can lead to addiction, depression, and other mental health issues.

Are there any medical uses for dissociative drugs?

Dissociative drugs have been used in medical settings to treat pain, anxiety, and other psychological disorders. However, they are highly regulated and should only be used under the supervision of a medical professional.

How can I prevent the misuse of dissociative drugs?

The best way to prevent the misuse of dissociative drugs is to educate yourself about the risks and dangers associated with them. If you or someone you know is using dissociative drugs, seek help from a medical professional.

In conclusion, dissociative drugs significantly impact individuals, society, and the world. Therefore, it is crucial to understand the effects and potential dangers. This book provides a comprehensive and engaging examination of this topic, making it an essential resource for anyone interested in learning about dissociative drugs.

1

KETAMINE

An unexpected journey down the Rabbit Hole as you experience the surreal and dreamlike effects of ketamine.

Ketamine, also known as special K, kit kat, cat valium, or simply K, is a powerful anesthetic that has been used medically for decades. Recently, it has gained notoriety for its use as a recreational drug. This chapter aims to provide a comprehensive overview of ketamine, covering its history, chemical properties, medical uses and abuses, risks and side effects, and laws and regulations. By presenting a well-rounded picture of this drug, we aim to give readers the knowledge they need to make informed decisions about ketamine use. Whether you are a curious reader, a medical professional, or someone considering using ketamine, this chapter has something for you. So sit back, relax, and join us as we explore the fascinating world of ketamine.

Historical Background

Ketamine was first synthesized in 1962 by the American pharmaceutical company Parke-Davis. It was developed as a safe, fast-acting anesthetic for **surgical procedures**, particularly when traditional anesthetics were unsuitable. Its unique properties, including its ability to produce a dissociative state in patients, quickly made it popular in medical circles.

In the 1970s and 1980s, ketamine was widely used in veterinary medicine. However, by the late 1980s, it had become a popular recreational drug. This led to increased regulations, restrictions on its use, and concerns about its potential for abuse. Despite this, ketamine continued to be used for medical purposes, particularly in emergency and trauma situations where its rapid onset and short duration of action were valuable.

In recent years, ketamine has seen a resurgence in popularity as a treatment for depression, anxiety, and other mental health conditions. This has led to renewed interest in its potential therapeutic benefits, as well as its potential risks and side effects. With a rich and complex history, ketamine remains one of the most intriguing substances in medicine and drug use.

Chemical Properties and Effects

Ketamine is a synthetic drug belonging to the arylcyclohexylamine class of drugs. It is a powerful anesthetic that acts on the central nervous system and can produce a dissociative state in users. When used medically, it is administered intravenously, intramuscularly, or orally.

The effects of ketamine can vary greatly depending on the dose and the method of administration. At low doses, it can produce mild, *trance-like* effects. In contrast, higher doses can cause hallucinations, dream-like states, and a feeling of detachment from the body and the environment. In addition, the dissociative state produced by ketamine can lead to a loss of consciousness. This property makes it a useful anesthetic in medical procedures.

In addition to its anesthetic properties, ketamine has been shown to have pain-relieving and anti-inflammatory effects, making it useful in managing chronic pain conditions. It is also used as an antidepressant and a treatment for anxiety and other mental health conditions due to its ability to reduce symptoms of depression and rapidly improve mood.

The precise mechanisms by which ketamine produces its effects are not yet fully understood. Still, it is thought to act on specific receptors in the brain that are involved in regulating mood and consciousness. Further research is needed to understand the chemical properties and effects of ketamine fully. Still, its unique combination of therapeutic and recreational effects makes it a substance of ongoing interest and study.

Medical Uses and Abuses

Ketamine is a widely used anesthetic in medical settings. It is prized for its fast onset, short duration of action, and low risk of causing respiratory depression. It is often used in emergency and trauma situations, as well as in children, due to its safety profile. In addition to its anesthetic properties, ketamine has also been used to manage chronic pain conditions and depression, anxiety, and other mental health conditions.

Despite its therapeutic uses, ketamine has also been abused for its hallucinogenic and dissociative effects. In addition, it has been used recreationally for its ability to produce intense and often unpredictable experiences. Its popularity as a party drug has led to concerns about its potential for abuse and addiction.

The abuse of ketamine can lead to various negative health effects, including hallucinations, confusion, paranoia, and impaired motor function. In addition, high doses can cause temporary paralysis, unconsciousness, and respiratory depression, leading to serious health consequences and even death.

To reduce the risk of abuse and negative health effects, ketamine is carefully regulated in many countries, and its use is restricted to medical settings. Despite this, it remains a widely used and abused substance, and efforts are ongoing better to understand its therapeutic benefits and potential for abuse.

Risks and Side Effects

Like all drugs, ketamine risks side effects and adverse reactions. The most common side effects of ketamine use include confusion, agitation, hallucinations, and impaired motor function. In addition, high doses can cause temporary paralysis, unconsciousness, and respiratory depression, leading to serious health consequences and even death.

Long-term use of ketamine can lead to various negative health effects, including bladder dysfunction, liver damage, and cognitive impairment. In addition, evidence suggests that repeated exposure to ketamine can lead to tolerance and dependence and persistent changes in brain function and behavior.

In addition to the physical risks associated with ketamine use, it can also have psychological effects, including anxiety, paranoia, and depression. These effects can be especially pronounced in individuals with a history of mental health conditions or who are particularly sensitive to the effects of drugs.

It is important to be aware of the risks and side effects associated with ketamine use and to use the drug only under the supervision of a medical professional. In addition, individuals considering using ketamine for recreational purposes should be aware of the risks involved and should take steps to minimize these risks and protect their health.

Laws and Regulations

Ketamine is a **controlled substance in many countries**, and its use is restricted to medical settings. The same laws and regulations surrounding ketamine vary by country and jurisdiction. Still, it is generally classified as a Schedule III controlled substance, meaning it is illegal to manufacture, distribute, or possess the drug for non-medical purposes.

In the United States, ketamine is regulated by the Drug Enforcement Administration (DEA) under the Controlled Substances Act. This classification makes it illegal to produce, distribute, or possess ketamine for non-medical purposes. Those caught doing so can face severe legal penalties, including fines and imprisonment.

In other countries, the laws and regulations surrounding ketamine may be even more restrictive. For example, the drug may be classified as a Schedule I or Schedule II controlled substance, with even greater penalties for those caught possessing or distributing the drug illegally.

Despite its strict legal status, ketamine remains a widely used and abused substance, and efforts are ongoing better to understand its therapeutic benefits and potential for abuse. To minimize the risk of negative health effects and legal consequences, it is important to be aware of the laws and regulations surrounding ketamine and to use the drug only under the supervision of a medical professional.

Experience Reports

Welcome to the experience reports section! Here, you can read first-hand accounts from people who have taken ketamine and experienced its effects. Their stories provide valuable insight into the wide range of experiences that can occur when taking this substance – both positive and negative. In addition, these reports will help you understand what to expect when taking ketamine.

. . .

Report 1: Ketamine Brings Celebratory Joy

I had heard about Ketamine before, but I was never brave enough to try it until I attended a party recently. I was interested to see what the experience would be like and was curious to find out. I figured why not try it and see what happens.

So, I took the plunge and consumed some Ketamine. At first I felt a little anxious and apprehensive, but soon the worries melted away and I found myself feeling a sense of joy and freedom. I felt my body move to the music in a way that I had never felt before. Every beat seemed to have its own unique rhythm, and I was one with the music and my body was euphorically dancing to its own beat. I felt a sense of connection with everyone around me, as if we were all part of a single energy. My face was stuck in a permanent smile, and I just felt so in tune with the present moment.

The experience was honestly amazing and unlike anything I had ever felt before. I felt so light and free, as if I was floating on a cloud. It was a truly uplifting and intense experience, and I am so glad that I had the courage to try something new. I would definitely recommend Ketamine to anyone who is looking for an unforgettable experience.

Report 2: A Negative Ketamine Experience

A few weeks ago, I decided to try out the drug ketamine. Everyone had told me it was a beautiful, enlightening experience, and I was interested to see for myself. However, when I took it, I was met with a completely different reality. Instead of feeling empowered and enlightened, I was left feeling confused, disoriented and muted. My ability to communicate and make sense of my surroundings had been taken away from me. I was unable to enjoy the experience, and instead, was filled with an intense sense of paranoia and fear. My body felt heavy and I felt like I had no control over myself, making the experience a very negative one for me. I was disappointed that my experience was so different from what I had hoped for, and it left me feeling overwhelmed.

Report 3: A Moment of Intense Creativity Through Ketamine

I had been struggling with a creative block for months, feeling like I was stuck in a rut and unable to move forward with any of the projects I had begun. But after trying ketamine, everything changed. I suddenly had a clarity of thought I hadn't experienced in months -- a clarity that allowed me to see my creative potential in a new light.

The first art piece I created after this experience was unlike any I had done before. I felt as though I had been freed from the chains of my own creativity and was now able to express myself in a way that was true and genuine. I felt inspired and motivated to keep creating, and soon found myself in a flow state I had never experienced before.

I soon realized that my newfound clarity wasn't just a temporary burst of creativity, but rather a permanent shift in my perception. I was now more aware of my own strengths and weaknesses, and had a newfound appreciation for the creative process. I was able to approach each project with a newfound confidence and enthusiasm, knowing that I could create something unique and powerful.

This newfound clarity stayed with me for months, allowing me to explore my creative potential and become a better artist. I was now able to create work that I was truly proud of, and I was constantly inspired and motivated to keep creating. I had found a peace and contentment within myself that I had never experienced before, and I was able to use this newfound knowledge to continue to develop my creative abilities.

My experience with ketamine was life-changing, and has allowed me to explore my creative potential and inner peace in ways that I never could have imagined before. I am forever grateful for this experience, and will continue to use the knowledge I have gained to help me reach new heights in my creative endeavors.

. . .

In conclusion, ketamine is a powerful anesthetic with a rich and complex history, ranging from its use in medical procedures to its notoriety as a recreational drug. Its chemical properties and effects on the central nervous system make it a unique substance of interest and study. However, it is important to be aware of the risks and side effects associated with its use, including confusion, agitation, hallucinations, and potentially serious health consequences. Despite its restrictions and regulations, ketamine remains a widely used and abused substance. Further research is needed to understand its therapeutic benefits and potential for abuse fully. Whether you are a curious reader, a medical professional, or considering using ketamine, it is essential to have a well-rounded understanding of this fascinating substance.

Chapter Summary

- Ketamine is a powerful anesthetic used medically for decades, but it is also abused as a recreational drug.
- It was first synthesized in 1962 by Parke-Davis as a safe, fast-acting anesthetic. It gained popularity in medical circles due to its unique properties and ability to produce a dissociative state in patients.
- In the late 1980s, ketamine became a popular recreational drug leading to increased regulations and restrictions on its use.
- Recently, ketamine has seen a resurgence in popularity as a treatment for depression, anxiety, and other mental health conditions.
- It is a synthetic drug belonging to the arylcyclohexylamine class. Its effects vary based on the dose and method of administration, ranging from mild trance-like effects to hallucinations and a feeling of detachment.
- The precise mechanisms of how ketamine produces its effects are not yet fully understood, but it is thought to act on specific

receptors in the brain involved in regulating mood and consciousness.

- Despite its therapeutic uses, ketamine is also abused for its hallucinogenic and dissociative effects. Its abuse can lead to negative health effects such as hallucinations, confusion, paranoia, impaired motor function, and even death.
- Its use is restricted to medical settings and carefully regulated. Still, it remains widely used and abused, and there is an ongoing effort to better understand its benefits and potential for abuse.

2

PCP (PHENCYCLIDINE)

The air was thick and heavy as a PCP-fueled haze blanketed the city, a surreal landscape of neon-tinged buildings and alleys. In the distance, a faint hum of bass and synthetic beats could be heard, setting the tone for a night of psychedelic exploration.

Phencyclidine, also known as PCP, angel dust, hog, or love boat, is a powerful and controversial substance with a rich history that spans decades. It is a dissociative anesthetic that was first developed in the 1950s as an intravenous anesthetic for surgical procedures but was eventually **abandoned** due to its unpredictable and sometimes dangerous side effects. Despite this, PCP has found its way into the recreational drug market, where it continues to be used for its mind-altering effects. In this chapter, we will explore the historical background of PCP, its chemical properties and effects, its medical uses and abuses, the risks and side effects associated with its use, the laws and regulations surrounding it, and finally, experience reports from those who have used it. Whether you are a curious reader or a student of drug culture, this chapter offers a comprehensive and engaging look at one of the most enigmatic substances of our time.

Historical Background

The history of Phencyclidine (PCP) dates back to the 1950s when it was first synthesized as an intravenous anesthetic for surgical procedures. Developed by Parke-Davis, a pharmaceutical company, PCP was marketed under the trade name Sernyl and was used as a general anesthetic until 1965. However, society discontinued its widespread use due to reports of unpredictable and dangerous side effects, including hallucinations, delusions, and agitation.

Despite its abandonment as a medical anesthetic, PCP found its way into the recreational drug market in the 1960s and 1970s. It gained a reputation as a potent hallucinogen. Its popularity grew; by the 1980s, PCP was one of the most commonly abused drugs in the United States. This led to increased media attention and a heightened public concern, leading to a crackdown on the production and sale of PCP.

Despite this, PCP remains a significant problem in the world of illegal drugs, and its use remains controversial. While it has been banned in most countries, PCP continues to be manufactured and sold on the black market, often under different names. The historical background of PCP is a story of a potent drug that has been celebrated and reviled. It serves as a cautionary tale of the dangers of recreational drug use.

Chemical Properties and Effects

Phencyclidine (PCP) is a dissociative anesthetic that disrupts communication between the brain and the body. It acts as a selective antagonist at the N-methyl-D-aspartate (NMDA) receptor, responsible for pain perception, learning, and memory.

Chemically, PCP is a white crystalline powder soluble in water and is often sprinkled on marijuana or tobacco. Individuals can take it orally or intravenously. The effects of PCP vary depending on the dose and the method of administration. Still, they can last anywhere from 4 to 24 hours.

At low doses, PCP can cause euphoria, empathy, and detachment from reality. Higher doses can cause symptoms such as hallucinations, delusions, agitation, and paranoia. PCP is also known to cause profound alterations in perception, including changes in body image and the distorted experience of time and space.

In addition to its mind-altering effects, PCP has several physiological effects, including increased blood pressure and heart rate, dry mouth, and decreased body temperature. Unfortunately, it can also cause muscular rigidity and seizures, which can be dangerous and even life-threatening in some cases.

In conclusion, the chemical properties and effects of PCP are complex and far-reaching, and they can profoundly impact the mind and body. Furthermore, the unpredictable nature of PCP and its potential for dangerous side effects highlight the importance of avoiding its use.

Medical Uses and Abuses

Phencyclidine (PCP) was first developed in the 1950s as an intravenous anesthetic for surgical procedures. Still, it was eventually **abandoned** due to its unpredictable and dangerous side effects. Despite its abandonment as a medical anesthetic, PCP found its way into the recreational drug market, which continues to be used for its mind-altering effects.

The use of PCP as a recreational drug has been associated with several negative outcomes, including hallucinations, delusions, and paranoia, as well as physical symptoms such as muscle rigidity, seizures, and an increased risk of overdose. Additionally, PCP is highly addictive, and long-term use can lead to permanent cognitive impairment and behavioral problems.

Despite its well-known risks and dangers, PCP continues to be used and abused by individuals seeking its powerful mind-altering effects. In addition, it is often used with other drugs, such as marijuana or alcohol, which can increase its potential for harm.

In conclusion, the medical uses and abuses of PCP are a testament to the dangers of recreational drug use. While PCP was once developed as a medical anesthetic, its unpredictable and dangerous side effects have led to its abandonment in that field. Its continued use as a recreational drug highlights the importance of avoiding drugs with unknown and potentially harmful effects.

Risks and Side Effects

The use of Phencyclidine (PCP) is associated with several serious risks and side effects, both physical and psychological. These can include:

- **Hallucinations and delusions:** PCP is known to cause profound alterations in perception, including hallucinations, delusions, and paranoia.
- **Agitation and violence:** PCP can cause aggression, anger, and violence, and it has been associated with several violent crimes.
- **Seizures and muscle rigidity:** PCP can cause muscle rigidity, convulsions, and seizures, which can be dangerous and even life-threatening in some cases.

- **Increased heart rate and blood pressure:** PCP can cause a rapid increase in heart rate and blood pressure, which can be particularly dangerous for individuals with heart conditions.
- **Addiction:** PCP is known to be highly addictive, and long-term use can lead to permanent cognitive impairment and behavioral problems.
- **Cognitive impairment:** PCP use can lead to memory loss, confusion, and other cognitive problems that can persist even after the drug has left the body.
- **Psychological distress:** PCP use can cause feelings of anxiety, depression, and other forms of psychological distress.

In conclusion, the risks and side effects of PCP are serious and far-reaching, and they highlight the importance of avoiding its use. Whether you are considering using PCP recreationally or have a history of drug abuse, it is important to understand the potential consequences of its use and to seek professional help if necessary.

Laws and Regulations

Phencyclidine (PCP) is a controlled substance in many countries, including the United States, where it is classified as a Schedule II drug under the Controlled Substances Act. This classification indicates that PCP has a high potential for abuse and limited medical use and is subject to strict controls and regulations.

In the United States, the possession, sale, and distribution of PCP are illegal, and individuals caught with the drug can face serious criminal charges, including imprisonment and fines. In addition, the manufacture and sale of PCP are also regulated, and individuals who engage in these activities can face additional legal consequences.

The laws and regulations surrounding PCP are intended to discourage its use and to minimize the harm associated with its abuse. By making PCP illegal and subject to criminal sanctions, authorities aim to reduce the drug's availability and encourage individuals to seek help if they have a problem with its use.

In conclusion, the laws and regulations surrounding PCP are an important component of efforts to reduce the harm associated with its use. By making PCP illegal and subject to criminal sanctions, authorities aim to discourage its use and minimize the harm associated with its abuse. Suppose you have questions or concerns about PCP or its regulation. In that case, it is important to seek the guidance of a knowledgeable professional.

Experience Reports

Welcome to the experience reports section! Here, you can read first-hand accounts from people who have taken PCP (Phencyclidine) and experienced its effects. Their stories provide valuable insight into the wide range of experiences that can occur when taking this substance – both positive and negative. In addition, these reports will help you understand what to expect when taking PCP.

Report 1: A Trip of a Lifetime

Before my experience with PCP, I had only heard about it from my friends and never actually tried it. I was hesitant and had some doubts, but I decided to take a chance.

On the day of my experience, I was feeling a range of emotions, but I was mostly excited. When I took a small dose, I felt an immediate rush of energy and enthusiasm.

Initially, I experienced a sense of weightlessness and my senses seemed to be heightened. I began exploring my environment and took in very vivid and intense details that I never noticed before. I also felt like my physical and mental barriers had been removed and I could explore without caring about my limitations.

The most vivid, and intense, experience was the intense connection to the music that played in the background. Every musical note seemed to communicate with me directly.

I had an overall sense of connection to the world around me, but also felt that I had a certain detachment from reality. I was happy but at the same time felt like I was somehow removed from everyday life.

By the end of my experience, I felt a great sense of relaxation combined with a feeling of heightened optimism. This mood of optimism enabled me to work through challenges in my life which had previously seemed insurmountable.

Overall, my experience with PCP was positive and unbelievably intense. I felt a connection to my environment in ways that I had never experienced before, and I felt a sense of optimism that enabled me to work through problems and issues in life. I believe my experience with Phencyclidine was one of a kind and life-changing, both mentally and emotionally.

Report 2: Damaging Effects of Phencyclidine

Unfortunately, I had the dubious honor of personally experiencing the dangers of taking Phencyclidine (PCP). Though I was aware of the risks of indulging in the substance, I underestimated the harm it could cause.

Upon taking the drug, I was immediately struck with a wave of dizziness and nausea. Despite this uncomfortable feeling, I stoically soldiered on in my endeavor until worse symptoms started to present. Mental confusion and paranoia began to set in, leaving me feeling anxious and disoriented. I was struck with a dangerous sense of detachment from reality that I had never before experienced. My bodily senses began to fail, and I noticed intense and disorganized sensory input.

My hallucinations intensified as the hours wore on. I felt as though I was being watched, although I was aware this was not the case. The acuity of my senses heightened exponentially, making the smallest noises and sensations unbearable. My vision felt distorted and obstructed, making it difficult to pilot myself through any environment. Nightmares followed, taking me to places of extreme fear and dread.

It became increasingly hard to concentrate. Additionally, the overwhelming feelings of paranoia and fear continued to haunt me. I felt completely out of control and could not trust myself or my thoughts.

In the long-term, the physical and mental effects of the drug have not worn off. My anxiety is higher than ever, and I've been plagued with insomnia since the experience. The event has left me feeling constantly uneasy and terrified that I will succumb to the effects of the drug again.

I can unequivocally say, as a warning to anyone who may consider taking Phencyclidine, that it is not a substance to trifle with. The risks far outweigh any reward, physical or mental, and the effects are often more powerful and longer-lasting than anticipated.

Report 3: Experiencing Euphoria

I recently had my first experience with Phencyclidine (PCP), and it was a beautiful and profound experience. After carefully considering the risks that come with any use of an illicit substance, I decided I was ready to take the plunge and try some PCP.

I could immediately feel the effects within just a few minutes. The physical effects were mild; I was able to maintain an upright posture and remain in control of my body for the duration of my experience. I felt a deep sense of peace and calm, and I was pleasantly surprised at how quickly and dramatically I could change my mental state, even at such low doses.

The visual effects were by far the most profound. I could feel my mind opening up as time and space seemed to melt together and everything around me took on a vivid and beautiful quality. Colors and sounds seemed to blend together in a way I had never seen before and my thoughts and emotions flowed through me in an incredibly profound and positive way.

The most wonderful part of my experience was the intense feeling of euphoria and love I felt for everything around me. I could feel and appreciate the beauty of the world in a completely new way and I was filled with a deep and meaningful understanding of the complexities of the universe.

This experience was easily one of the most profound and transformative experiences of my life. I have a newfound appreciation and understanding of the world around us and the limitless possibilities that exist within our consciousness.

In conclusion, PCP, commonly known as Phencyclidine, is a potent and controversial substance with a rich history. Despite being developed in the 1950s as an intravenous anesthetic, it was eventually abandoned due to its unpredictable and dangerous side effects. Nevertheless, PCP found its way into the recreational drug market, where it continues to be used for its mind-altering effects. The historical background, chemical properties and effects, medical uses and abuses, and risks and side effects associated with PCP serve as a cautionary tale of the dangers of recreational drug use. Therefore, it is important to avoid PCP and drugs with unknown and potentially harmful effects and seek help for substance abuse issues.

Chapter Summary

- PCP (Phencyclidine) is a dissociative anesthetic developed in the 1950s for surgical procedures but was later abandoned due to its unpredictable side effects.
- PCP has a rich history and remains a controversial substance despite its dangers.
- PCP acts as a selective antagonist at the NMDA receptor and causes profound alterations in perception, including hallucinations and delusions.
- PCP is a white crystalline powder that can be taken orally or intravenously and can cause physiological effects like increased blood pressure and heart rate, muscle rigidity, and seizures.

- PCP was once used as a medical anesthetic but has since been banned in most countries and is now used as a recreational drug.
- The use of PCP as a recreational drug has been associated with negative outcomes such as hallucinations, delusions, and violence, as well as a high risk of addiction and long-term cognitive impairment.
- PCP is often used with other drugs, such as marijuana or alcohol, which can increase its potential for harm.
- PCP is associated with serious risks and side effects, including hallucinations, delusions, violence, seizures, muscle rigidity, and long-term cognitive impairment.

3

DXM (DEXTROMETHORPHAN)

As the sun sets on the horizon, a young man takes a deep breath and embraces the euphoria of his first DXM experience.

Dextromethorphan, also known as DXM is a **versatile substance** with a complex history and a range of effects. Originating as a cough suppressant in over-the-counter cold and flu medications, DXM has gained attention for its potential for abuse and illicit use as a recreational drug. This chapter will explore the historical background of DXM, its chemical properties and effects, medical uses and abuses, risks, and side effects, laws and regulations, and experience reports. Whether you're a medical professional, a curious individual, or someone looking for a comprehensive understanding of DXM, this chapter aims to provide an insightful and comprehensive guide to this multifaceted substance.

Historical Background

The history of Dextromethorphan, or DXM, dates back to the 1950s when it was first synthesized and developed as a cough suppressant. It was initially marketed in cough syrups and other over-the-counter cold and flu medications. As a result, its use as a cough suppressant became widespread. However, our society did not widely recognize its potential for abuse as a recreational drug until the late 20th century, when reports of individuals intentionally consuming large quantities of cough syrup for its psychoactive effects began to emerge.

In the decades since, DXM has become a widely abused substance, particularly among adolescents and young adults. In response to this trend, many countries have enacted laws and regulations restricting access to cough syrups containing DXM. In addition, some countries classify it as a controlled substance. Despite these measures, DXM abuse remains a significant public health concern. Consequently, it continues to be the subject of ongoing research and analysis.

Overall, the historical background of DXM reflects its journey from a simple cough suppressant to a widely abused recreational drug with a complex and evolving regulatory landscape. Its legacy continues to shape the perception and understanding of DXM. It serves as a reminder of the need for continued vigilance and awareness regarding substance abuse.

Chemical Properties and Effects

Dextromethorphan, or DXM, is a synthetic substance with unique chemical properties that give rise to its various effects. It is classified as a morphinan and acts as a non-competitive antagonist of the N-methyl-D-aspartate (NMDA) receptor in the central nervous system. This interaction results in the suppression of coughing and the alteration of perceptions and mood, making it both a cough suppressant and a recreational drug.

DXM is considered safe and effective for its intended use as a **cough suppressant**. However, higher doses can produce psychoactive effects, including euphoria, changes in perception, dissociation, and altered thinking and behavior. In addition, the specific effects of DXM can vary depending on the dose, individual tolerance, and other factors. Still, it is considered a dissociative substance.

It is important to note that the effects of DXM can be unpredictable, and its use carries significant risks, particularly when taken in large doses. Moreover, DXM can interact with other substances, including alcohol and drugs, increasing the risk of adverse effects and potentially life-threatening consequences.

In summary, the chemical properties of DXM play a significant role in determining its therapeutic and recreational effects. Therefore, understanding these properties is crucial to safely and effectively using this substance and avoiding its associated risks.

Medical Uses and Abuses

Dextromethorphan, or DXM, has both medical and recreational uses. It is widely available in over-the-counter cold and flu medications as a cough suppressant. In addition, it is used to treat symptoms of coughs, colds, and respiratory infections. DXM is considered safe and effective at recommended doses for its intended use.

However, the same properties that make DXM an effective cough suppressant make it attractive for abuse as a recreational drug. At high doses, DXM can produce psychoactive effects, including euphoria, changes in perception, and dissociation, leading some individuals to intentionally consume large quantities of cough syrup or other products containing DXM for its psychoactive effects. This use pattern is considered abuse and carries significant risks, including physical and psychological harm.

The abuse of DXM has been a growing concern in recent decades. It is particularly prevalent among adolescents and young adults. As a result, many countries have enacted laws and regulations to restrict access to cough syrups and other DXM products to address this trend, with some countries classifying it as a controlled substance.

In conclusion, while DXM has medical uses as a cough suppressant, its abuse as a recreational drug remains a significant public health concern. Therefore, it is important to understand the risks associated with DXM abuse and only use it as directed by a healthcare provider. Individuals should also be aware of the laws and regulations surrounding DXM and take steps to protect themselves and others from its harmful effects.

Risks and Side Effects

Dextromethorphan, or DXM, has both therapeutic and recreational uses. Still, its use carries significant risks and potential side effects, particularly in high doses. At low doses, DXM is considered safe for its intended use as a cough suppressant. Still, its abuse as a recreational drug can lead to various adverse effects, including physical and psychological harm.

Some of the most common side effects of DXM abuse include nausea and vomiting, dizziness and drowsiness, increased heart rate, and blood pressure, and changes in perception and thinking. High doses can also cause more severe side effects, such as hallucinations, agitation, confusion, seizures, and impaired coordination.

In addition to the physical side effects of DXM abuse, significant psychological risks are associated with its use. These risks can include feelings of anxiety and paranoia, depression, and long-term changes in mood and behavior. In addition, the dissociative effects of DXM can also lead to feelings of detachment and disorientation. They may increase the risk of accidents or harm oneself or others.

DXM can also interact with other substances, including alcohol and drugs, increasing the risk of adverse effects and potentially life-threatening consequences. It is also important to note that the long-term effects of DXM abuse are not well understood, and more research is needed to fully understand the risks and consequences associated with this pattern of use.

In conclusion, the risks and side effects associated with DXM use, both therapeutic and recreational, highlight the importance of using this substance only as directed by a healthcare provider and avoiding its abuse. Individuals should also be aware of the potential for adverse effects and take steps to minimize the risk of harm.

Laws and Regulations

The laws and regulations surrounding Dextromethorphan, or DXM, vary depending on the jurisdiction. Still, in many countries, it is subject to legal controls and restrictions to prevent abuse. This is due to the growing concern about the harmful effects associated with DXM abuse, particularly among adolescents and young adults.

In some countries, DXM is classified as a controlled substance. Its sale and distribution are subject to strict regulations, including age restrictions and prescription or over-the-counter availability requirements. In other countries, it is classified as a legal but regulated substance. Its sale and distribution may be subject to restrictions or limitations.

It is important to note that the laws and regulations surrounding DXM can change over time and vary from country to country. Therefore, individuals should be aware of the legal status of DXM in their jurisdiction and comply with all relevant laws and regulations regarding its use and abuse.

In addition to national laws and regulations, some international agreements may also play a role in controlling and regulating DXM. For example, the United Nations Convention on Psychotropic Substances requires signatory countries to control and regulate the production, distribution, and use of certain psychoactive substances, including DXM.

In conclusion, the laws and regulations surrounding DXM reflect the growing concern about its abuse and the associated risks. They are designed to minimize the harm associated with this substance. It is important for individuals to be aware of the legal status of DXM in their jurisdiction and to comply with all relevant laws and regulations regarding its use and abuse.

Experience Reports

Welcome to the experience reports section! Here, you can read first-hand accounts from people who have taken DXM (Dextromethorphan) and experienced its effects. Their stories provide valuable insight into the wide range of experiences that can occur when taking this substance – both positive and negative. In addition, these reports will help you understand what to expect when taking DXM.

Report 1: Reflecting on a Euphoric Experience with DXM

I recently had the opportunity to try DXM, and I can confidently say that it was an experience like no other. From the moment I took a dose of the substance, I found myself in a state of peace and tranquility that was both calming and uplifting.

I was amazed at the intense rush of euphoria that washed over me, feeling like I was disconnected from time and reality. All of my anxieties and worries seemed to float away, and in the place of them came feelings of joy and contentment. The colors around me looked brighter, and the music I was listening to sounded sweeter.

The visuals I experienced during this trip were also breathtaking. I saw vibrant colors and intricate patterns that filled my field of vision. For hours I felt like I could just sit and appreciate the beauty of it all.

In addition to the euphoria, DXM also gave me the mental clarity to more deeply reflect on life's big questions. I felt like I was able to make some extraordinary insights about my life and the world around me.

I was surprised to find that the effects of DXM lasted longer than I expected and that the comedown felt incredibly smooth. Despite the intensity of the experience, I felt well-rested, level headed, and energized when it was done.

In summary, DXM was an extremely positive experience for me. It not only left me feeling joyous and content, but I also gained a lot from it in terms of personal insights. I would highly recommend it to those looking to explore a unique state of consciousness.

Report 2: A Disaster of a DXM Trip

It all started out as harmless curiosity. I heard from a friend that DXM, an over-the counter cough syrup, could be used as a recreational drug to achieve a high. Naive as I was, I decided to try it for myself.

Regretfully, my DXM experience was nothing close to pleasant. As soon as I swallowed the syrup, feelings of anxiety and restlessness began to set in. My heart rate accelerated, my body became shaky, and my mind began to warp. I attempted to get comfortable by laying down, but my head began to race with streams of seemingly disconnected thoughts and ideas. I started to feel incredibly out of control and disoriented.

The physical effects of the DXM were awful—I felt extremely nauseous and dizzy, and I began vomiting uncontrollably. The sensation felt like an extended version of a car ride that's taken a wrong turn down a bumpy road. As the feeling intensified, I thought I was going to faint multiple times, and I was having difficulty thinking straight.

It would have been easy to just press pause, but the power of the drug seemed to cast a spell over me that I couldn't break. My trip seemed like it would never end, and every single thought felt like an eternity.

Now that it's been some time since my DXM experience, I'm glad it's finally over. I realized that day that DXM is not for me. It left me feeling empty, degraded, and scared. Overall, it was a disaster and a lesson learned.

Report 3: Riding the DXM Express

I recently decided to explore the depths of my psyche by experimenting with DXM (Dextromethorphan). My journey began with a measured dose of a DXM-containing cough syrup. As the viscous syrup trickled down my throat, I was beginning to get excited for the adventure that lay ahead.

The effects of the DXM started off slow and gentle, as I began to feel a pleasant tingling and warmth radiating throughout my body. Everything began to take on a dreamlike quality, as if I were living an alternate reality separate from the one I know. I felt immersed in an ever-shifting collage of colors and shapes, with my imagination running wild and my attention fixed firmly on the present.

My senses were heightened and I became acutely aware of the subtleties of sound and touch that usually go unnoticed. I experienced closeness to the people around me, regardless of whether or not I knew them and regardless of our differences. I had a sense of connectedness to all living beings.

The DXM had made me aware of the beauty in my own backyard and provided a new perspective on the world around me. I felt light, yet grounded; enlightened, yet grounded in the present moment.

The effects of the DXM continued to build and deepen as I became more and more introspective. I gained insight into myself and my relationships that I had not seen before. I finally felt at peace and ready to end my journey.

All in all, DXM was a unique and fascinating experience. It allowed me to explore my inner world and discover things about myself I had never known before. I am now more aware of my surroundings, my emotions, and my relationships with those around me. I have embraced the new perspective that DXM has given me and have come away from the experience feeling ready to take on life's challenges.

In conclusion, Dextromethorphan, or DXM, is a complex substance with a versatile range of effects, originating as a cough suppressant and later gaining attention as a recreational drug. This chapter provided an in-depth exploration of the historical background, chemical properties and effects, medical uses and abuses, risks, and side effects of DXM. It is important to understand the properties of this substance to safely and effectively use it and avoid its associated risks. While DXM has therapeutic uses, its abuse as a recreational drug remains a **significant public health concern**. Therefore, individuals should be aware of the laws and regulations surrounding it. Additionally, the use of DXM carries significant risks and potential side effects, and it is crucial only to use it as directed by a healthcare provider.

Chapter Summary

- DXM, also known as Dextromethorphan, is a versatile substance with a complex history, originally developed as a cough suppressant in over-the-counter cold and flu medications.
- Despite its widespread use as a cough suppressant, DXM has gained attention for its potential for abuse and illicit use as a recreational drug.
- DXM is classified as a morphinan and acts as a non-competitive antagonist of the N-methyl-D-aspartate (NMDA) receptor in the central nervous system, suppressing coughing and alterations in perceptions and mood.
- The effects of DXM can be unpredictable, and its use carries significant risks, particularly when taken in large doses or combined with other substances.

- DXM abuse has become a growing public health concern, particularly among adolescents and young adults, leading to regulations restricting access to cough syrups containing DXM in many countries.
- DXM has medical and recreational uses, with its abuse as a recreational drug carrying significant physical and psychological harm.
- The side effects of DXM abuse can range from nausea and vomiting, dizziness, increased heart rate, and blood pressure to more severe effects such as hallucinations, confusion, seizures, and impaired coordination.
- Understanding the risks and potential harm associated with DXM abuse is important for this substance's responsible and safe use.

4

NITROUS OXIDE (LAUGHING GAS)

A bright and sunny day at the park, with a rainbow of balloons floating in the sky, each one a unique and beautiful color.

Nitrous oxide, also known as laughing gas, balloons, chargers, hippie crack, or nangs, is a colorless and odorless gas that has captured people's interest for centuries. From its early use as a recreational substance in the 18th century to its widespread use in modern medicine, nitrous oxide has played a unique role in science, medicine, and recreation. This chapter will delve into the fascinating history of nitrous oxide, its chemical properties, effects, and medical applications, as well as the risks and side effects associated with its use. We will also explore the laws and regulations governing nitrous oxide and review some experience reports to better understand the recreational use of the gas. So join us on a journey through nitrous oxide's history, science, and cultural impact.

Historical Background

The history of nitrous oxide dates back to the late 18th century when English chemist and inventor Joseph Priestley first discovered the gas. However, the English chemist and natural philosopher Sir Humphry Davy recognized nitrous oxide's potential as a recreational substance. In the late 1700s and early 1800s, nitrous oxide parties, or "laughing gas parties," became popular in Europe and the United States. During these parties, people would inhale nitrous oxide for its euphoric and analgesic effects.

In addition to its recreational use, nitrous oxide has significantly impacted the field of medicine. In the late 19th century, American dentist Horace Wells used nitrous oxide as an anesthetic during dental procedures, marking the first recorded use of nitrous oxide in surgery. Over the next few decades, nitrous oxide became widely used in medicine, particularly in dentistry and surgery, due to its fast-acting and short-lived effects.

Today, nitrous oxide continues to be used as a medicinal and recreational substance. Yet, despite its long history, nitrous oxide still holds many **mysteries** and is the subject of ongoing scientific research and exploration.

Chemical Properties and Effects

Nitrous oxide, also known as dinitrogen oxide, is a colorless and odorless gas composed of nitrogen and oxygen. It has the chemical formula N2O and is stored as a compressed liquid in steel cylinders. Nitrous oxide is a powerful anesthetic and a weak psychoactive substance with both physical and mental effects.

When inhaled, nitrous oxide rapidly diffuses into the bloodstream. It produces a range of effects, including increased feelings of euphoria, happiness, and excitement. It also reduces pain perception, creates a sense of detachment, and produces hallucinations and altered states of consciousness. The duration of nitrous oxide's effects depends on the dose and the method of administration but typically lasts between 5 and 20 minutes.

In addition to its psychoactive effects, nitrous oxide has several physical effects. It acts as a powerful analgesic, reducing pain perception and allowing surgical procedures without causing significant patient discomfort. Nitrous oxide also produces mild sedation and has a calming effect. It is useful for reducing anxiety and fear in medical and dental procedures.

It is important to note that nitrous oxide has a low therapeutic index, meaning that the difference between a therapeutic dose and a lethal dose is small. As a result, nitrous oxide must be used with caution and under the supervision of a trained medical professional.

Medical Uses and Abuses

Nitrous oxide has been used in medicine for over a century. It is widely recognized for its safe and effective use as an anesthetic. In the medical field, nitrous oxide is primarily used in dental procedures, such as extractions, fillings, and cleanings, as well as in surgical procedures to provide sedation and pain relief. In addition, nitrous oxide's fast-acting and short-lived effects make it an ideal choice for medical procedures requiring minimal sedation.

However, nitrous oxide also has a history of abuse, particularly in the recreational context. As a result, people inhale nitrous oxide for its psychoactive effects risk serious injury or death and long-term damage to their health. In addition, abusing nitrous oxide can cause a range of negative effects, including brain damage, nerve damage, and vitamin B12 deficiency, which can lead to peripheral neuropathy.

In addition, nitrous oxide abuse can lead to addiction, as users may become dependent on the drug to experience its psychoactive effects. The inhalation of nitrous oxide in large quantities can also lead to unconsciousness and, in some cases, death.

It is important to note that nitrous oxide is regulated in many countries and is only available for medical use with a prescription. Misuse or abuse of nitrous oxide is a serious crime and can result in severe legal consequences.

Risks and Side Effects

Like all substances, nitrous oxide has associated risks and side effects ranging from mild to severe. Some of the nitrous oxide's most common side effects include dizziness, headache, nausea, and vomiting. Individuals can sometimes manage these side effects with proper medical supervision. Still, in other cases, they may indicate a more serious problem.

One of the major risks associated with nitrous oxide is the risk of overdose. Overdose can occur when a person inhales too much nitrous oxide, leading to unconsciousness and, in some cases, death. Additionally, inhaling nitrous oxide can reduce the amount of oxygen in the body, leading to brain damage and other serious health problems.

Long-term nitrous oxide abuse can also lead to serious health problems, such as nerve damage, B12 deficiency, and peripheral neuropathy. Chronic abuse of nitrous oxide can also cause changes in the brain, leading to depression, anxiety, and other mental health problems.

In addition, nitrous oxide abuse can lead to addiction, as users may become dependent on the drug to experience its psychoactive effects. Withdrawal from nitrous oxide can lead to symptoms such as headaches, nausea, and fatigue and may require medical intervention.

It is important to use nitrous oxide only as directed by a trained medical professional and to never inhale nitrous oxide directly from a canister or other source. Misuse or abuse of nitrous oxide can result in serious health problems and should be avoided.

Laws and Regulations

Nitrous oxide is regulated by the government in many countries and is classified as a controlled substance in some jurisdictions. Therefore, the possession, sale, and distribution of nitrous oxide without a valid prescription are illegal and can result in serious legal consequences, including fines and imprisonment.

In the medical field, nitrous oxide is only used by trained healthcare professionals. It is subject to strict regulations regarding storage, handling, and administration. Medical facilities that use nitrous oxide must follow strict protocols to ensure the safe and proper use of the substance.

In recreational terms, nitrous oxide is often sold on the black market. It is commonly used at music festivals, parties, and other events. The sale and distribution of nitrous oxide for recreational purposes are illegal. They can result in severe legal consequences, including fines and imprisonment.

It is important to be aware of nitrous oxide laws and regulations and only use the substance as directed by a trained medical professional. Misuse or abuse of nitrous oxide can have serious legal consequences and should be avoided.

Experience Reports

Welcome to the experience reports section! Here, you can read first-hand accounts from people who have taken nitrous oxide and experienced its effects. Their stories provide valuable insight into the wide range of experiences that can occur when taking this substance – both positive and negative. In addition, these reports will help you understand what to expect when taking nitrous oxide.

. . .

Report 1: My Wild Journey with Nitrous Oxide

Recently, I decided to try something new - nitrous oxide. To be honest, I was a bit scared, but I am so glad I took the plunge!

I had a wonderful experience with nitrous oxide. Immediately upon inhaling it, my senses were heightened, and I felt an incredible sense of euphoria. All my worries were gone and I was free to explore my own world. The sounds were clearer, and the colors were brighter than ever before. I even felt a bit weightless, like I was floating on a cloud. It was an awe-inspiring experience, and the beautiful state of relaxation I was in lasted for hours.

Nitrous oxide made me feel incredibly peaceful and connected to the world around me.

Report 2: Dizzy and Disoriented

Today, I decided to try something new and experiment with nitrous oxide. It was a chore to get my hands on it, but I eventually found a source. After inhaling the nitrous oxide, I waited for the effects to be felt.

I can't say I was impressed. I immediately felt dizzy and out of it. I almost felt like I was going to faint, and I felt really off balance. My vision went blurry and I had trouble staying focused on the task at hand.

Maybe the worst part of it was the lack of control. I felt like a puppet, and the substance was the person pulling the strings. I had no control over my own body, and my mind felt fuzzier than usual.

Unfortunately, this experience did not go as planned. I have tried many substances and thought nitrous oxide was fine to try too. However, the lack of control and dizziness it caused was more than I bargained for. If anyone else out there was considering nitrous oxide, I do not recommend it.

. . .

Report 3: A Painless Experience

I had been curious about nitrous oxide for a long time, and I finally had the opportunity to experience it. I felt a bit apprehensive at first, but my worries were soon replaced by excitement and anticipation.

The setup was straightforward; I was in a comfortable chair with a breathing mask over my nose. It was easy to find a suitable breathing rhythm without giving it much thought. After a few minutes, I began to feel a wave of intense warmth and lightness of being. It felt like the room around me was spinning ever so slightly, and I felt a newfound sense of euphoria, like I was living in a dream.

The sensation didn't last very long; it felt like only a few minutes, but it was a great sensation. I felt energetic and relaxed, but also still in control. Physically, I felt pretty refreshed and relaxed.

I had no qualms with the experience, and I'd recommend it to anyone looking for a relaxing experience without the worry of any side effects. I feel lucky that I was able to experience nitrous oxide, and I'll be doing it again in the future.

In conclusion, nitrous oxide, also known as laughing gas, has played a unique role in science, medicine, and recreation for centuries. From its discovery by Joseph Priestly and recognition of its potential by Sir Humphry Davy to its widespread use in modern medicine, nitrous oxide has left a lasting impact on history. Its chemical properties, including its composition and effects, have made it a valuable tool in medicine as a powerful anesthetic. However, nitrous oxide also has risks and side effects associated with its use, especially when abused. Therefore, it is important to use nitrous oxide with caution and under the supervision of a trained professional, as misuse can lead to serious injury or death, addiction, and legal consequences.

Chapter Summary

- Nitrous oxide, also known as laughing gas, has a fascinating history dating back to the late 18th century.
- Nitrous oxide is a colorless and odorless gas composed of nitrogen and oxygen, used as a powerful anesthetic and weak psychoactive substance.
- When inhaled, nitrous oxide produces euphoria, happiness, reduced pain perception, detachment, hallucinations, and altered states of consciousness.
- Nitrous oxide has been used in medicine for over a century, mainly in dental and surgical procedures.
- Abusing nitrous oxide has serious risks, including brain damage, nerve damage, B12 deficiency, addiction, unconsciousness, and death.
- Misuse of nitrous oxide is a serious crime regulated by many countries.
- Nitrous oxide has common side effects, such as dizziness, headache, nausea, and vomiting.
- Chronic nitrous oxide abuse can lead to serious health problems, including nerve damage, B12 deficiency, peripheral neuropathy, depression, and anxiety.

5

MXE (METHOXETAMINE)

A surreal landscape where MXE molecules are the only inhabitants, floating in a sea of neon colors.

Methoxetamine, also known as MXE, M-Ket, Max, or Mexxy, is a novel psychoactive substance recently gaining popularity. Despite its growing use, many people need to become more familiar with MXE and its effects. This chapter will delve into the history and background of MXE, its chemical properties, and its various medical uses and abuses. We will also examine its use's risks and side effects and the laws and regulations governing it. Finally, we will provide a collection of experience reports from individuals who have used MXE, offering a glimpse into the personal experiences of those who have used the substance. Whether you are a researcher, healthcare professional, or simply curious about MXE, this chapter aims to provide you with a comprehensive understanding of this fascinating substance.

Historical Background

The historical background of Methoxetamine, commonly known as MXE, is relatively brief. MXE was first synthesized in the **early 2010s** and quickly gained popularity as a recreational drug due to its supposed similar effects to the dissociative anesthetic ketamine. It was first sold online as a research chemical and marketed as a legal alternative to ketamine.

However, due to concerns over its safety and potential for abuse, many countries have since classified MXE as a controlled substance. Despite this, it remains readily available for purchase on the dark web and in some countries where it is not yet banned.

MXE has gained significant attention from the media and law enforcement, with reports of serious adverse events associated with its use, including hospitalizations and death. This highlights the need for further research into the effects and potential dangers of MXE and other similar substances.

It is important to note that the limited research available on MXE suggests that its long-term effects still need to be fully understood and may be much more potent and dangerous than originally thought. As a result, individuals are highly recommended to avoid using MXE and seek safer, more researched alternatives for recreational or therapeutic purposes.

Chemical Properties and Effects

MXE, or Methoxetamine, is a synthetic compound classified as a dissociative anesthetic. This means that it produces a state of detachment or dissociation from one's surroundings and physical sensations, leading to a feeling of floating and disconnection.

MXE has a chemical structure similar to ketamine. Its effects are thought to be caused by its actions on the glutamate and dopamine neurotransmitter systems. It is also believed to bind to the NMDA receptor in the brain, leading to its anesthetic and dissociative effects.

The effects of MXE can last anywhere from a few hours to several days and include altered perception of time and space, changes in body sensations, feelings of euphoria and detachment, and hallucinations. It can also lead to confusion, agitation, and intense hallucinations, known as a "bad trip." At high doses, MXE can cause unconsciousness, which can be dangerous and lead to accidental injury.

It is important to note that due to the limited research on MXE, its effects, and potency must be better understood. The substance can be unpredictable and potentially dangerous, and we should approach its use cautiously. Suppose you are considering using MXE or any other psychoactive substance. In that case, it is recommended that you speak with a healthcare professional and carefully research its effects, risks, and potential dangers.

Medical Uses and Abuses

MXE, or Methoxetamine, has not been approved for medical use and is not recognized as a safe or effective treatment for any medical condition. However, it has been used recreationally and abused for its dissociative and hallucinogenic effects.

Despite its lack of medical approval, some individuals have claimed that MXE has helped them with depression, anxiety, and other mental health conditions. However, these claims have not been scientifically proven, and using MXE for medical purposes is not recommended.

The recreational use of MXE is associated with several potential risks, including psychological dependence, tolerance, and withdrawal. Additionally, using MXE can lead to serious adverse events, such as accidental injury, overdose, and death.

MXE abuse has also been associated with negative physical and psychological effects, including confusion, agitation, panic attacks, delusions, and hallucinations. It can also lead to dangerous health consequences like cardiac and respiratory problems, seizures, and kidney failure.

It is important to note that using MXE, whether for medical or recreational purposes, carries significant risks and potential dangers. Therefore, if you are considering using MXE or any other psychoactive substance, it is recommended that you speak with a healthcare professional and carefully consider the potential risks and benefits. Abusing any substance can have serious consequences for your health and well-being and should be avoided.

Risks and Side Effects

MXE, or Methoxetamine, is associated with several potential risks and side effects. These can range from short-term effects, such as confusion and agitation, to more serious and long-lasting health problems. Some of the common short-term side effects of MXE include:

- Altered perception of time and space
- Changes in body sensations
- Confusion and disorientation
- Nausea and vomiting
- Agitation and panic attacks
- Hallucinations

The long-term effects of MXE use are not well understood, but some potential consequences include the following:

- Psychological dependence and tolerance
- Withdrawal symptoms, including anxiety and depression
- Cognitive impairments, such as memory problems and difficulty concentrating
- Respiratory and cardiovascular problems
- Kidney and liver damage
- Seizures and other neurological problems

It is important to note that using MXE can be **unpredictable** and lead to serious adverse events, including overdose and death. Therefore, if you are experiencing any negative effects from using MXE, it is recommended that you seek medical attention immediately.

Furthermore, MXE is not regulated, and its quality and purity can vary greatly, making it difficult to know what you are taking. This increases the risk of accidental overdose and serious health problems.

In conclusion, MXE, like any other psychoactive substance, carries significant risks and potential dangers. If you are considering using MXE, it is recommended that you speak with a healthcare professional and carefully weigh the potential risks and benefits. It is always safer to avoid using untested and unregulated substances.

Laws and Regulations

MXE, or Methoxetamine, is a relatively new substance whose legal status varies from country to country. In many countries, including the United States, the United Kingdom, and Australia, MXE is classified as a controlled substance, and its sale, possession, and use are illegal.

In the United States, MXE is classified as a Schedule III controlled substance, which means that it is considered to have a high potential for abuse and dependence and is illegal to manufacture, distribute, or possess.

MXE is classified as a Class B drug in the United Kingdom, and its sale, possession, and use are illegal. The penalties for violating the law regarding MXE can be severe, including fines and imprisonment.

In Australia, MXE is classified as a Schedule 9 substance, which means that it has a high potential for abuse and dependence and is illegal to manufacture, distribute, or possess.

It is important to note that the legal status of MXE can change rapidly, and it is recommended that you check the current laws and regulations in your jurisdiction before using or possessing the substance.

In conclusion, using and possessing MXE is illegal in many countries, and the penalties for violating the law can be severe. Therefore, if you are considering using MXE or any other psychoactive substance, it is recommended that you research the current laws and regulations in your jurisdiction and consult with a legal professional if necessary.

Experience Reports

Welcome to the experience reports section! Here, you can read first-hand accounts from people who have taken MXE (Methoxetamine) and experienced its effects. Their stories provide valuable insight into the wide range of experiences that can occur when taking this substance – both positive and negative. In addition, these reports will help you understand what to expect when taking MXE.

Report 1: A Blissful and Enlightening Journey with MXE

I recently had the experience of trying MXE, also known as methoxetamine. I settled into a comfortable chair, making sure I had all that I needed close by in case I wanted to move or shift my position.

The effects of the MXE slowly started to come on within about thirty minutes of taking it. I felt a noticeable shift in my mood, in addition to slight tingling sensations throughout my body. As the MXE fully began to take effect, I felt myself being taken away on an incredibly blissful journey. I experienced complete relaxation, a feeling of total euphoria, and a deep and profound connection to my inner being.

The MXE seemed to stimulate a deep contemplation within me, as I contemplated my life and the meaning of existence. I felt a true sense of peace and understanding, as I was able to look at life from a different perspective. A message of unconditional love resonated with me at the most profound levels.

As I moved towards the end of the MXE experience, I felt open and connected to my surroundings once more. I felt a deep sense of joy and freedom, as my insight and understanding of life had been completely altered for the better. Overall, the experience with MXE was incredibly peaceful and enlightening, and I am so grateful for the opportunity to have experienced it.

Report 2: My Horrible MXE Experience

I recently tried MXE, also known as methoxetamine, for the first time and it was an experience I never want to repeat. Going into it, I heard a lot of positive things about the drug, so I figured it sounded relatively harmless. Boy, was I wrong.

I took the drug and at first, I felt a sense of euphoria. However, the good feelings quickly vanished and were replaced with an intense rush of anxiety and paranoia. I felt an overwhelming sense of dread and my entire body felt completely overwhelmed. I went from feeling calm and relaxed to feeling total terror in a matter of moments.

The physical effects of MXE were equally unpleasant. I felt nauseous and lightheaded and I couldn't keep my balance. My vision blurred and every sound I heard felt like a million knives stabbing through my brain. My heart raced and felt like it was about ready to stop and I thought for sure I was going to go into cardiac arrest.

After what felt like an eternity, the effects of the drug finally wore off and I was left exhausted, drained and feeling worse than I ever had in my entire life. The hangover lasted for days afterwards, making it that much worse.

I regret ever trying MXE and would never recommend it to anyone. Avoid it at all costs!

. . .

Report 3: An Unexpected and Positive Psychedelic Experience with MXE

I recently had the opportunity to experiment with MXE for the first time and it was an unforgettable experience. After ingesting it, I began to feel a deep sense of relaxation, both mentally and physically. I felt like I was in a trance and could hear music in the background as if it was playing from within my own self.

As time passed, I began to drift away to a place of beautiful imagery. Colors and shapes began to appear, spinning and dancing in front of me. I felt as if I had transcended to a higher level of consciousness and was able to see things in a completely new way.

The effects of MXE lasted for several hours before slowly beginning to subside. As I gradually returned to my everyday consciousness, I was left with an overwhelmingly positive feeling. I felt inspired and energized, with a newfound clarity of mind.

For someone looking for a light, positive psychedelic experience, I would certainly recommend giving MXE a try. It opened my mind up to a fascinating new world of possibilities, and I'm grateful I had the chance to explore it.

In conclusion, Methoxetamine, also known as MXE, is a novel psychoactive substance that has recently gained popularity but remains largely misunderstood. Despite its growing use, it is important to understand its history, chemical properties, and medical uses and abuses. MXE is classified as a dissociative anesthetic and has similar effects to ketamine. Still, its long-term effects are not fully understood and may be much more potent and dangerous. The recreational use of MXE is associated with several potential risks and side effects, including psychological dependence, tolerance, withdrawal, and serious adverse events like accidental injury, overdose, and death. Therefore, individuals are highly recommended to avoid using MXE and seek safer, more researched alternatives. Before using any psychoactive substance, it is important to speak with a healthcare professional and carefully consider the potential risks and benefits.

Chapter Summary

- MXE, also known as Methoxetamine, is a synthetic dissociative anesthetic that has gained popularity as a recreational drug.
- MXE was first synthesized in the early 2010s and was marketed as a legal alternative to ketamine.
- MXE has been classified as a controlled substance in many countries due to concerns over its safety and potential for abuse.
- MXE produces a state of detachment or dissociation from one's surroundings and physical sensations, leading to a feeling of floating and disconnection.
- MXE has not been approved for medical use, and its potential risks and side effects still need to be fully understood.
- The recreational use of MXE is associated with several potential risks, including psychological dependence, tolerance, and withdrawal.
- Abusing MXE can lead to serious adverse events, such as accidental injury, overdose, and death, as well as negative physical and psychological effects.
- Using MXE carries significant risks and potential dangers and should be approached with caution. Speak with a healthcare professional before using MXE or any other psychoactive substance.

6

3-MEO-PCP

A young man lies in a state of euphoria, surrounded by clouds of pcp-induced smoke, while a rainbow of colors dance around him.

In recent years, 3-MeO-PCP, also known as 3-Methoxy-Phencyclidine, has gained attention due to its growing popularity as a recreational drug. This chapter will overview the historical background, chemical properties, and effects of 3-MeO-PCP. Additionally, we will delve into this substance's medical uses and abuses and the risks and side effects associated with its consumption. This chapter will also discuss the laws and regulations surrounding 3-MeO-PCP and personal accounts or experience reports from those who have used the substance. By the end of this chapter, readers will have a comprehensive understanding of 3-MeO-PCP and its impact on society.

Historical Background

The historical background of 3-MeO-PCP dates back to the late 1950s when phencyclidine (PCP) was first synthesized and developed as an anesthetic. However, due to its unpredictable and often dangerous side effects, it was discontinued for medical use. Instead, it gained notoriety as a recreational drug in the 1970s and 1980s.

In recent years, new derivatives of PCP have emerged, including 3-MeO-PCP. This substance has gained popularity among drug users for its hallucinogenic effects, which are reported to be similar to those of PCP, but with a shorter duration. Although 3-MeO-PCP is not as widely known as other recreational drugs, its use is growing. Unfortunately, it has been associated with many negative consequences, including overdoses and hospitalizations.

Despite its relatively short history as a recreational drug, 3-MeO-PCP has already established a reputation for being dangerous and unpredictable. As the use of this substance continues to increase, it is important to understand its effects and potential consequences, as well as its historical background and development.

Chemical Properties and Effects

3-MeO-PCP is a synthetic compound that belongs to a class of drugs known as dissociative anesthetics. It disrupts communication between the brain and body, resulting in feelings of detachment, altered perception, and, sometimes, hallucinations.

The chemical structure of 3-MeO-PCP is similar to that of PCP but with a few key differences. These modifications to the molecular structure have been reported to have different effects. For example, some users describe 3-MeO-PCP as having a more euphoric and less intense dissociative experience than PCP.

The effects of 3-MeO-PCP vary depending on the dose, method of administration, and individual characteristics of the user. Low doses can produce mild stimulation, euphoria, and mild hallucinations. Higher doses can cause intense **dissociation**, sedation, and significant alterations in perception and thinking.

It is important to note that 3-MeO-PCP is a potent and highly unpredictable substance. As a result, its effects can vary widely, even among experienced users. This unpredictability and the drug's tendency to produce intense and sometimes frightening experiences make it a particularly dangerous substance to use recreationally.

Medical Uses and Abuses

3-MeO-PCP was never approved for medical use and had no accepted medical applications. Despite this, some people have sought out the drug for its dissociative effects and potential to produce hallucinations.

The recreational use of 3-MeO-PCP has become increasingly popular recently, particularly among those seeking a more intense or hallucinogenic experience. The drug is typically consumed orally, although some users may snort or inject it.

The abuse of 3-MeO-PCP can have serious consequences, including overdose, hospitalization, and long-term health problems. In addition, the effects of the drug can be highly unpredictable, even among experienced users. They can include extreme anxiety, agitation, and violent behavior.

In addition to its potential to cause physical harm, 3-MeO-PCP can also have significant psychological effects, including hallucinations, delusions, and paranoia. These effects can be particularly dangerous for individuals with pre-existing mental health conditions, as the drug can exacerbate their symptoms and lead to a more severe reaction.

Given the serious risks associated with the abuse of 3-MeO-PCP, individuals need to be aware of the dangers of using this substance and avoid it altogether. Furthermore, healthcare providers and substance abuse specialists should be prepared to address the issue of 3-MeO-PCP abuse and its consequences as the use of this drug continue to increase.

Risks and Side Effects

Using 3-MeO-PCP can carry significant risks and side effects in the short and long term. Some of the most common side effects include:

- **Dissociation:** This is the most pronounced effect of 3-MeO-PCP and can result in feelings of detachment from reality, altered perceptions of time and space, and, sometimes, hallucinations.
- **Agitation and anxiety:** The drug can cause intense agitation and anxiety, particularly at higher doses, leading to dangerous and unpredictable behavior.
- **Physical symptoms:** 3-MeO-PCP can cause many physical symptoms, including high blood pressure, heart palpitations, muscle rigidity, and seizures.
- **Cognitive impairment:** The drug can cause significant cognitive impairment, including confusion, memory loss, and difficulty with problem-solving and decision-making.
- **Overdose:** 3-MeO-PCP is a potent drug; even a small overdose can result in serious consequences, including coma and death.

In addition to the short-term risks, 3-MeO-PCP use can lead to long-term health problems. For example, frequent drug use can cause tolerance, increase the risk of overdose, and cause lasting damage to the brain and other organs.

Given these serious risks and side effects, individuals need to be aware of the dangers of using 3-MeO-PCP and avoid it altogether. Suppose you or someone you know is struggling with substance abuse. In that case, it is important to seek help from a healthcare provider or substance abuse specialist.

Laws and Regulations

3-MeO-PCP is a controlled substance in many countries, and its possession, sale, and distribution are strictly regulated. The same laws and regulations surrounding 3-MeO-PCP vary from country to country. Still, in general, the drug is classified as a Schedule I or Schedule II controlled substance, meaning that it is considered to have a high potential for abuse and no accepted medical use.

In the United States, 3-MeO-PCP is classified as a Schedule I controlled substance under the Controlled Substances Act. This means that it is illegal to manufacture, distribute, or possess the drug, and penalties for violating these laws can be severe, including fines and imprisonment.

In the European Union, 3-MeO-PCP is regulated by the European Monitoring Centre for Drugs and Drug Addiction (EMCDDA) and is classified as a controlled substance in most member states.

Given the strict laws and regulations surrounding 3-MeO-PCP, individuals must be aware of the legal consequences of using and possessing the drug. Suppose you have questions about your area's laws and regulations surrounding 3-MeO-PCP. In that case, it is best to consult a knowledgeable attorney or legal expert.

Experience Reports

Welcome to the experience reports section! Here, you can read first-hand accounts from people who have taken 3-MeO-PCP and experienced its effects. Their stories provide valuable insight into the wide range of experiences that can occur when taking this substance – both positive and negative. In addition, these reports will help you understand what to expect when taking 3-MeO-PCP.

Report 1: The Most Amazing Insight into Reality

When I decided to try 3-MeO-PCP, I had no idea how much I'd be rewarded. This trip was truly remarkable.

The experience began slowly - I felt visual distortions that slowly became more intense. I experienced more depth and variety of color than ever before and I felt that my surroundings had taken on an entirely new life. I found myself smiling more and more as the effects became stronger, like I was in a secret and sacred place that no one else knew.

The peak of the experience turned out to be even more remarkable. I felt incredibly connected with my inner self and the universe around me. I felt like I understood life in a way I could never have imagined. Everything I saw and experienced felt so real, so perfect, and it made me feel so much love for the people and things around me. I came away from the experience with a heightened understanding of life and the way I wanted to live it.

Overall, my experience with 3-MeO-PCP was truly remarkable. I feel that it has opened me up to new ways of looking at the world and I'm grateful for the insights I've gained through this amazing experience.

Report 2: No More

I can't believe I did it. I'm so disappointed in myself that I actually had the audacity to try 3-MeO-PCP. I had heard that it was supposed to be a powerful psychedelic and a dissociative, but I had no idea just how intense the experience would be.

I had thought that I was prepared to handle its psychedelic effects, but boy was I wrong. As soon as I took it, I immediately felt the urge to vomit, and my heart started pounding out of my chest. I felt like I was having an out-of-body experience as my sense of reality slowly started to slip away.

I attempted to hold onto my sense of reality with every ounce of strength that I had, but it was too late. I was mentally, emotionally, and physically overloaded by the 3-MeO-PCP. I had the worst trip of my life, and my mental state was in shambles when it eventually began to wear off.

Never again will I make the mistake of taking 3-MeO-PCP. This experience was far too intense, and I'm not sure if I'll ever be able to fully recover from it. My advice to anyone considering trying 3-MeO-PCP is to think twice before doing so. It is an incredibly powerful substance and should not be taken lightly.

Report 3: My Trippy Trip to the Third Dimension with 3-MeO-PCP

I recently had the pleasure of exploring the otherworldly splendor that 3-MeO-PCP had to offer. After carefully reading up and researching the potential effects, I felt prepared to embark on my own psychedelic journey.

I snuck away to a quiet and comfortable area, where I could feel safe and secure while I "took off". This was paramount to the experience, as I felt it allowed myself to be at ease and not be opposed to the expedition I was about to embark on.

I took the 3-MeO-PCP in the form of an insufflated dose to get the most intense effects, and the results were beyond my expectations. A wave of euphoria engulfed me, and I had a greater connection to the world around me. I was still aware of what was going on, but visions and insights presented themselves to me, and I felt as though I was exploring a brightly colored, fully-immersive world, so vivid and intense it verged on the unreal. I was transported to the beyond, into a world that I would never forget.

When the effects started to wear away, I relaxed, feeling mesmerized but contented. I returned to the real world with newfound facial expressions, a spiritual connection and a greater appreciation for life. I had certainly explored a new dimension, one that I plan to visit again and again.

In conclusion, 3-MeO-PCP, a synthetic compound, and dissociative anesthetic, has gained popularity as a recreational drug due to its hallucinogenic effects. However, its use has been associated with negative consequences such as overdoses, hospitalizations, and long-term health problems. This substance is highly unpredictable, and its effects can vary widely, even among experienced users, leading to potentially dangerous situations. Furthermore, the drug has no accepted medical applications. It can cause significant psychological harm, particularly in individuals with pre-existing mental health conditions. Individuals must be aware of the dangers of using 3-MeO-PCP and avoid it altogether. Healthcare providers and substance abuse specialists must be prepared to address the issue as its use continues to increase.

Chapter Summary

- 3-MeO-PCP, also known as 3-Methoxy-Phencyclidine, is a growingly popular recreational drug with hallucinogenic effects.
- The historical background of 3-MeO-PCP dates back to the late 1950s when it was first synthesized and developed as an anesthetic but discontinued due to its dangerous side effects.
- 3-MeO-PCP is a synthetic compound that belongs to a class of drugs known as dissociative anesthetics, disrupting communication between the brain and body.

- The effects of 3-MeO-PCP vary with dose, method of administration, and individual characteristics and can range from mild stimulation to intense dissociation and alterations in perception.
- 3-MeO-PCP was never approved for medical use and had no accepted medical applications.
- The abuse of 3-MeO-PCP can result in serious consequences such as overdose, hospitalization, long-term health problems, and worsening mental health conditions.
- The side effects of 3-MeO-PCP use include dissociation, agitation and anxiety, physical symptoms, cognitive impairment, and overdose.
- 3-MeO-PCP use carries significant short and long-term risks, and individuals need to be aware of its dangers and avoid it. In addition, healthcare providers and substance abuse specialists must be prepared to address the issue of its abuse.

7

SALVIA DIVINORUM

A woman enters a dark cave, illuminated only by the moonlight that streams from a hole in the ceiling. Inside, she finds a strange plant with purple flowers - Salvia Divinorum.

Salvia divinorum, also known as Maria Pastora, Sally-D, or simply Salvia, is a powerful psychoactive plant used for centuries for its mind-altering effects. With its growing popularity, it's essential to understand the historical background, chemical properties, medical uses, risks, and laws surrounding this plant. In this chapter, we will dive into salvia divinorum and explore its impact on human consciousness. From its ancient origins to its modern-day use, we'll cover everything you need to know about this intriguing substance. We'll also touch on the potential medical benefits and the possible dangers associated with its use. Lastly, we'll examine salvia's legal landscape and share some personal experience reports to give you a glimpse into what to expect. Whether you're a seasoned salvia user or just curious about this unique plant, this chapter offers something.

Historical Background

The historical background of salvia divinorum dates back to the indigenous people of Mexico, who have used the plant for its psychoactive effects for centuries. The Mazatec indigenous tribe, who lived in the Oaxaca region of Mexico, were the first to discover the plant's mystical properties. They used salvia divinorum in traditional shamanic healing ceremonies, considered a powerful tool for inducing spiritual visions.

The plant was largely unknown to the Western world until the **mid-20th century** when anthropologists and botanists first learned of its use by the Mazatec tribe. It was later introduced to the Western world in the 1950s and 1960s, quickly gaining popularity among psychonauts and those seeking a new and unique mind-altering experience.

In the early 2000s, salvia divinorum started gaining notoriety as a recreational drug. Its popularity spread rapidly, especially among young people. This led to increased concerns about its safety and potential risks, which prompted many countries to regulate its use and sale.

Today, salvia divinorum continues to be a subject of fascination, with new research being conducted on its effects and potential medical uses. Despite the controversies surrounding its use, the plant remains an important part of cultural and spiritual traditions for many people. Moreover, its historical background is a testament to its lasting impact on human consciousness.

Chemical Properties and Effects

The active ingredient in salvia divinorum is salvinorin A, a highly potent and selective kappa opioid receptor agonist. This means that when ingested, salvinorin A binds to the kappa opioid receptors in the brain and produces a range of effects, including alterations in perception, thought, mood, and consciousness.

The effects of salvia divinorum are highly variable. They can range from mild to intense, depending on the dose and method of ingestion. When taken in low doses, the plant can produce mild effects such as a mild buzz or feeling of relaxation. However, when taken in higher doses, the effects can be much more intense and can last anywhere from 15 minutes to an hour.

Some of the most commonly reported effects of salvia divinorum include changes in perception, such as visual and auditory hallucinations, feelings of detachment from the body, and intense euphoria or anxiety. The plant can also cause alterations in thought patterns, leading to a profound change in one's perspective or worldview.

It is important to note that the effects of salvia divinorum can be highly unpredictable and can vary greatly from person to person. This variability makes it difficult to predict the exact nature and intensity of the effects. It underscores the importance of caution and responsible use when experimenting with this substance.

Medical Uses and Abuses

Salvia divinorum has been studied for its potential medical uses. Although the research is still in its early stages, there is evidence to suggest that the plant may have therapeutic potential. Some of the most promising areas of research include the treatment of addiction, depression, and pain.

For example, studies have shown that salvia divinorum has the potential as a treatment for opioid addiction. The plant's active ingredient, salvinorin A, has been shown to reduce cravings and withdrawal symptoms in people addicted to opioids. Additionally, the plant's ability to induce feelings of detachment from the body and alter perception may make it useful as a therapeutic tool for people struggling with depression and anxiety.

Despite these potential benefits, it is important to note that salvia divinorum for medicinal purposes still needs to be better understood. More research is needed to understand its effects and potential risks fully.

On the other hand, the plant can also be abused, particularly as a recreational drug. Its intense and highly variable effects can lead to dangerous and unpredictable outcomes, particularly when taken in large doses or combined with other substances. Additionally, the plant's potency and rapid onset of action can lead to accidental overdoses, which can be dangerous and even life-threatening.

Therefore, it is crucial to approach salvia divinorum with **caution** and use it only under the supervision of a trained medical professional. Furthermore, individuals should carefully balance the plant's potential benefits against its potential risks, and the latest scientific evidence should guide its use.

Risks and Side Effects

Salvia divinorum is a potent psychoactive plant that can significantly affect the mind and body. While the plant has been used for centuries for its medicinal and spiritual properties, it also carries some risks and side effects that individuals must consider.

Some of the most common side effects of salvia divinorum include nausea, dizziness, headache, loss of coordination, and impaired judgment. Additionally, the plant's intense and unpredictable effects can be dangerous, particularly for people with a history of mental health problems, as it can trigger psychotic symptoms or exacerbate existing conditions.

In some cases, using salvia divinorum can lead to accidents or dangerous behaviors, such as driving while under the influence or engaging in dangerous activities while under the influence. This highlights the importance of using caution and avoiding heavy operating machinery or engaging in any other activities that require alertness and good judgment while under the influence of the plant.

Additionally, the plant's potency and rapid onset of action can lead to accidental overdoses, which can be dangerous and even life-threatening. Therefore, it is important to be aware of the potential risks associated with using salvia divinorum and approach its use with caution, particularly for people who are sensitive to its effects or have a history of mental health problems.

In conclusion, while salvia divinorum has potential therapeutic benefits, it is important to be aware of its potential risks and side effects. Therefore, its use should be guided by the latest scientific evidence and only under the supervision of a trained medical professional.

Laws and Regulations

The legal status of salvia divinorum varies widely across the world, with some countries regulating or outright banning its use. In contrast, others allow it to be sold and consumed freely.

In the United States, salvia divinorum is a controlled substance in several states, meaning its sale, possession, and use are restricted by law. However, the plant is unregulated in other states and can be purchased and used freely. At the federal level, salvia divinorum is not a controlled substance. Still, the Drug Enforcement Administration (DEA) has listed it as a "Drug of Concern."

In Europe, the legal status of salvia divinorum also varies widely from country to country. Some countries, such as the United Kingdom, regulate its use, while others, such as the Netherlands, allow its use without restriction.

It is important to note that the legal status of salvia divinorum can change rapidly and is subject to various interpretations and regulations. Therefore, it is essential to stay informed about the latest laws and regulations regarding its use in your specific jurisdiction and to be mindful of the potential legal consequences of using this substance.

In conclusion, the legal status of the salvia divinorum is complex and rapidly evolving, with different countries and jurisdictions adopting different approaches to regulating its use. Therefore, before using this substance, it is important to be aware of the laws and regulations in your area and to use it only in compliance with the latest legal requirements.

Experience Reports

Welcome to the experience reports section! Here, you can read first-hand accounts from people who have taken salvia divinorum and experienced its effects. Their stories provide valuable insight into the wide range of experiences that can occur when taking this substance – both positive and negative. In addition, these reports will help you understand what to expect when taking salvia divinorum.

Report 1: My Amazing Salvia Divinorum Trip

I recently had the pleasure of trying salvia divinorum for the first time, and the experience was nothing short of remarkable. The intense physical sensations and changes in my perception of reality were the most unusual sensations I've ever experienced.

The trip began with me ingesting some of the salvia and sitting down in a comfortable chair, prepared for what I was about to experience. At first, I felt a faint tingling throughout my whole body, and a feeling of lightheadedness as though I was entering a daze. The sensations progressed rapidly, and before I knew it, I was falling through a never-ending tunnel. I felt a strange sense of freedom and detachment, which seemed to be an upside-down form of meditation.

As I ventured further, I noticed my environment had changed, and I was now walking through surreal forests, lush meadows, and curious structures made from random objects. Colors seemed to glow from everything, and the music I heard was changing from moment to moment, morphing from classical tunes to different genres of electronic music. It was unlike anything I'd ever heard before.

At some point, I became aware of a small creature standing in front of me. It looked like an elf or a fairy, but I couldn't really tell what it was. I felt an immediate connection with it, as if we had known each other for centuries. We talked for a while and I asked it questions about the universe. It gave me advice and shared its wisdom with me. After a while, I began to feel as though I was slowly coming back to Earth, or at least a version of it.

My salvia divinorum experience was an incredible one. I'm so glad that I had the chance to explore this amazing plant medicine and uncover its many mysteries. It was both a fascinating and enlightening journey, which taught me a lot about life and the universe. Without a doubt, it was one of the most incredible trips of my lifetime.

Report 2: A Destructive Experience

Recently, I decided to experiment with the powerful natural psychedelic known as salvia divinorum. After doing some research, I felt perhaps I had a grasp of what I was getting into. However, nothing could have prepared me for this utterly terrifying experience.

From the moment I ingested the small amount of salvia divinorum, I felt a strange and overwhelming sense of dread. Over the course of the next hour, my anxiety only built as things around me seemed to become warped and distorted. Colors were more intense and I experienced a frightening sense of detachment from reality. As if I observing a bizarre dream, I was unable to physically control my own limbs or any movements.

I had heard that this experience can be overwhelming and even uncomfortable, but I had no idea it could be so destructive. I felt as though my mind was being broken down and split into different pieces. I could no longer trust any of my senses, and I was in a constant state of fear and confusion.

The effects did eventually wear off and I am now safely back to reality. I feel as though I am still trying to make sense of everything that happened, but in hindsight, I can confidently say that salvia divinorum is not for me. I highly suggest that others abstain from experimenting with this drug, as the experience was far from pleasant.

Report 3: A Tale of Illuminating Inspiration

I recently had the opportunity to try out salvia Divinorum as a means of achieving personal enlightenment. It's a type of sage that is native to Mexico and is known to enhance mental clarity and inspire creativity.

I had heard rumors of people claiming to experience tremendous spiritual journeys under its influence, so I was extremely excited to see what it could do for me. The afternoon of my experiment, I set up my room with some gentle music, candles, and plenty of pillows for ultimate comfort.

When I finally took the plunge and tried it for the first time, I was amazed to find out how powerful this leaf was! I experienced a surge of strange visuals and audible hallucinations that lasted for about fifteen minutes. During this span of time, I explored never-before-imagined realms and received great insight into some of the more complicated issues in my life.

The result of my experience was undeniably positive. I was able to gain insight into areas of my life that had been difficult to pinpoint and make sense of before. Additionally, I felt inspired and became motivated to pursue the dreams and goals that I had been neglecting for some time.

Overall, using salvia Divinorum left me feeling deeply inspired and loved. It was a wonderful opportunity to explore my innermost thoughts, and a great reminder that there is much more to life than the mundane. I couldn't speak more highly of this experience, and would recommend it to anyone looking for an enlightening journey.

. . .

In conclusion, salvia divinorum is a powerful psychoactive plant with a rich historical background that has been used for centuries by the indigenous people of Mexico for its mind-altering effects. Its active ingredient, salvinorin A, binds to the kappa opioid receptors in the brain and produces a range of effects, including alterations in perception, thought, mood, and consciousness. While research is ongoing, there is evidence to suggest that salvia divinorum may have therapeutic potential, particularly in treating addiction, depression, and pain. However, it is important to approach its use with caution and to consider its potential risks and side effects, including nausea, dizziness, impaired judgment, and unpredictable effects. Overall, it is crucial to use salvia divinorum only under the supervision of a trained medical professional and to balance its potential benefits against its potential risks in light of the latest scientific evidence.

Chapter Summary

- Salvia divinorum is a powerful psychoactive plant used by the indigenous people of Mexico for centuries.
- The active ingredient in salvia is salvinorin A, a potent and selective kappa opioid receptor agonist.
- The effects of salvia can vary greatly, ranging from mild to intense, and can include alterations in perception, thought, mood, and consciousness.
- Salvia has potential medical benefits, including treating addiction, depression, and pain.
- Salvia also has the potential to be abused, and individuals should approach its use with caution.
- Common side effects of salvia include nausea, dizziness, headache, loss of coordination, and impaired judgment.
- The effects of salvia can be unpredictable and dangerous, especially for individuals with a history of mental health problems.

- The use of salvia should be guided by caution and the latest scientific evidence.

8

DISSOCIATIVE DRUGS AND MENTAL HEALTH

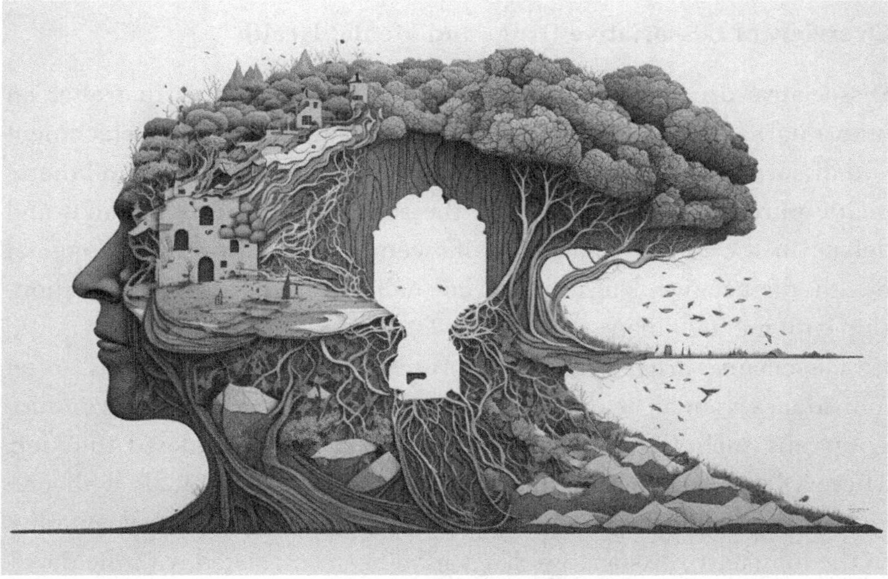

A surreal landscape that captures the complexity of mental health.

Dissociative drugs have long been a topic of interest in scientific and popular communities. These drugs can alter an individual's perception of reality and have been used for recreational and therapeutic purposes. The relationship between dissociative drugs and mental health is complex and multifaceted. This chapter will explore the relationship between dissociative drugs and mental illness, the symptoms these drugs may induce or exacerbate, and their potential role in treating mental illness. However, it's also important to examine the potential dangers of using dissociative drugs as a treatment, as they can have serious and long-lasting effects on one's mental and physical well-being. Through thoroughly examining the latest research and expert opinions, this chapter aims to provide a comprehensive overview of the relationship between dissociative drugs and mental health.

Overview of Dissociative Drugs and Mental Health

Dissociative drugs are a class of psychoactive substances that alter an individual's perception of reality and can induce feelings of detachment and dissociation. These drugs have been used for recreational and therapeutic purposes. They have been the subject of ongoing research and debate in the medical and scientific communities. In terms of mental health, dissociative drugs have been found to have a complex relationship with mental illness.

Dissociative drugs can have short-term and long-term effects on an individual's mental health. In the short term, these drugs can induce symptoms such as hallucinations, delusions, and disordered thinking. These effects can be particularly pronounced in individuals with pre-existing mental health conditions and can exacerbate existing symptoms. In the long term, dissociative drugs have been associated with the development of persistent symptoms of mental illness, such as depression, anxiety, and paranoia.

It's also important to note that dissociative drugs can have physical health consequences, such as heart palpitations, muscle spasms, and tremors. Moreover, they can have serious and long-lasting effects on an individual's brain function, including memory impairment and cognitive difficulties.

The relationship between dissociative drugs and mental health is complex and multifaceted. Therefore, it's essential to continue researching this topic and understanding the potential consequences of these drugs, both positive and negative, to provide informed and effective treatment for individuals with mental illness.

The Relationship Between Dissociative Drugs and Mental Illness

The relationship between dissociative drugs and mental illness is a complex and controversial issue that has been the subject of ongoing research and debate in the medical and scientific communities. On the one hand, some individuals with mental illness may use dissociative drugs as a form of self-medication to relieve symptoms of their condition. On the other hand, dissociative drugs can have negative consequences. It can exacerbate existing symptoms or even lead to new mental health problems.

Studies have shown that individuals with certain mental illnesses, such as depression, anxiety, and post-traumatic stress disorder (PTSD), may be more likely to use dissociative drugs as a form of self-medication. This is because dissociative drugs can induce feelings of detachment and dissociation, providing temporary relief from mental illness symptoms.

However, dissociative drugs can also lead to the development of persistent and persistent symptoms of mental illness, such as depression, anxiety, and paranoia. In some cases, long-term use of dissociative drugs can even lead to new mental health problems, such as psychotic disorders.

It's also important to note that individuals with pre-existing mental health conditions are at a higher risk of experiencing adverse effects from dissociative drugs. These effects can include increased symptoms of anxiety, depression, and paranoia, as well as an increased risk of self-harm and suicide.

In conclusion, the relationship between dissociative drugs and mental illness is complex and multi-faceted. While some individuals may use dissociative drugs as a form of self-medication, these drugs can have serious and long-lasting consequences on one's mental and physical well-being. Therefore, it's essential to continue researching this topic to provide informed and effective treatment for individuals with mental illness.

Dissociative Drugs and Symptoms of Mental Illness

Dissociative drugs have been found to have a complex relationship with mental illness, and their use can induce both short-term and long-term symptoms of mental illness. In the short term, dissociative drugs can induce symptoms such as hallucinations, delusions, and disordered thinking. These symptoms can be particularly pronounced in individuals with pre-existing mental health conditions and can exacerbate existing symptoms.

In the long term, dissociative drugs have been associated with the development of persistent symptoms of mental illness, such as depression, anxiety, and paranoia. In addition, studies have shown that individuals with pre-existing mental health conditions are at a higher risk of experiencing these symptoms after using dissociative drugs.

It's also important to note that dissociative drugs can lead to new mental health problems, such as psychotic disorders. In addition, in some cases, individuals may experience symptoms of mental illness even after they have stopped using dissociative drugs.

In addition to inducing symptoms of mental illness, dissociative drugs can also harm an individual's overall functioning. This can include memory and cognitive functioning difficulties, decreased motivation, and decreased ability to engage in daily activities.

In conclusion, dissociative drugs have been found to have a complex relationship with mental illness. They can induce both short-term and long-term symptoms of mental illness. In addition, using these drugs can also lead to new mental health problems and harm an individual's overall functioning. Therefore, it's essential to continue researching this topic to understand the relationship between dissociative drugs and mental illness and provide effective treatment for individuals affected by these conditions.

The Role of Dissociative Drugs in Treating Mental Illness

The role of dissociative drugs in treating mental illness is a controversial and complex issue that has been the subject of ongoing research and debate in the medical and scientific communities. While some studies have found that dissociative drugs may have therapeutic benefits for individuals with certain mental illnesses, such as post-traumatic stress disorder (PTSD) and depression, the potential dangers and negative consequences associated with the use of these drugs must be taken into consideration.

In the treatment of PTSD, for example, dissociative drugs have been found to have a calming effect on the central nervous system, reducing anxiety and fear. This can provide temporary relief from symptoms of PTSD, allowing individuals to engage in therapy and other forms of treatment.

However, it's important to note that the use of dissociative drugs in treating mental illness must be carefully monitored and used under the guidance of a medical professional. This is because the long-term effects of these drugs on an individual's mental and physical well-being are not yet fully understood, and individuals with pre-existing mental health conditions are at a higher risk of experiencing adverse effects.

In conclusion, the role of dissociative drugs in treating mental illness is a complex issue that requires ongoing research and careful consideration. While dissociative drugs may have therapeutic benefits for some individuals with certain mental illnesses, the potential dangers and negative consequences associated with the use of these drugs must be taken into account when making treatment decisions. Therefore, individuals with mental illness should only use dissociative drugs under the guidance of a medical professional and with a comprehensive and individualized treatment plan in place.

The Potential Dangers of Using Dissociative Drugs to Treat Mental Illness

The use of dissociative drugs to treat mental illness is associated with a range of potential dangers, including both short-term and long-term risks to an individual's mental and physical health.

In the short term, dissociative drugs can induce symptoms such as hallucinations, delusions, and disordered thinking. These symptoms can be particularly pronounced in individuals with pre-existing mental health conditions and can exacerbate existing symptoms. In addition, dissociative drugs can also result in impaired motor coordination and an increased risk of accidents, particularly in individuals who engage in activities such as driving or operating machinery.

Long-term use of dissociative drugs can lead to persistent symptoms of mental illness, such as depression, anxiety, and paranoia. Additionally, individuals who use dissociative drugs over an extended period are at a higher risk of developing a substance use disorder, such as addiction. This can lead to further negative consequences, including financial difficulties, social problems, and decreased overall functioning.

Furthermore, dissociative drugs can also hurt an individual's overall well-being, including difficulties with memory and cognitive functioning and decreased motivation and ability to engage in daily activities.

It's also important to note that the long-term effects of dissociative drugs on an individual's mental and physical health are not fully understood, and ongoing research is necessary to determine the full extent of these risks.

In conclusion, using dissociative drugs to treat mental illness is associated with various potential dangers, including short-term and long-term risks to an individual's mental and physical health. Therefore, individuals with mental illness who are considering dissociative drugs as a form of treatment should only use these drugs under the guidance of a medical professional and with a comprehensive and individualized treatment plan in place.

Chapter Summary

- Dissociative drugs alter an individual's perception of reality and have been used for recreational and therapeutic purposes.
- The relationship between dissociative drugs and mental health is complex and multifaceted.
- Dissociative drugs can have short-term and long-term effects on mental health, including inducing symptoms of mental illness and exacerbating existing symptoms.
- Dissociative drugs can also lead to new mental health problems and physical health consequences.
- The relationship between dissociative drugs and mental illness is controversial, with some individuals using them as a form of self-medication.
- Individuals with pre-existing mental health conditions are at a higher risk of experiencing adverse effects from dissociative drugs.
- Dissociative drugs can harm an individual's overall functioning and lead to memory and cognitive difficulties.
- Further research is necessary to understand the relationship between dissociative drugs and mental health to provide informed and effective treatment.

9

DISSOCIATIVE DRUGS AND SOCIETY

As the sun set, the cityscape was illuminated in a kaleidoscope of colors, reflecting the diversity and complexity of modern society.

Dissociative drugs have been a topic of **concern** for many years, affecting the individuals who use them and society. In this chapter, we will delve into the various impacts of dissociative drugs on crime and violence, the economy, healthcare, education, and more. We will also examine the government and law enforcement's role in addressing the use and abuse of dissociative drugs. With a clear and concise approach, this chapter aims to provide a comprehensive understanding of the complex issue of dissociative drugs and their impact on society.

Overview of Dissociative Drugs and Society

Dissociative drugs, also known as dissociatives, are a class of psychoactive substances that produce feelings of detachment and disconnection from one's surroundings and self. Some most commonly known dissociative drugs include ketamine, PCP, and DXM.

Dissociative drugs have been a topic of concern for society due to the potential harm they can cause to individuals and the wider community. These drugs can cause various negative effects, including hallucinations, confusion, agitation, and impaired motor function. High doses can also lead to serious health issues such as respiratory failure, coma, and even death.

Dissociative drugs have also been linked to increased crime and violence. In some cases, individuals under the influence of dissociatives have been involved in violent crimes, putting themselves and others at risk. The impact of dissociative drugs on the economy cannot be ignored either, as the cost of healthcare, law enforcement, and criminal justice resources used to address the problem is substantial.

In conclusion, dissociative drugs have significant and far-reaching effects on society. Therefore, it is important for individuals, communities, and the government to be aware of the dangers associated with these drugs and to work together to address the issue.

The Impact of Dissociative Drugs on Crime and Violence

The impact of dissociative drugs on crime and violence is a major concern for society. These drugs can induce confusion, agitation, and violence in users, leading to dangerous and unpredictable behavior. In some cases, individuals under the influence of dissociatives have committed violent crimes, putting themselves and others at risk.

Studies have shown that dissociative drugs are associated with an increased likelihood of criminal behavior, including assault, robbery, and domestic violence. For example, ketamine has been linked to violent behavior, and individuals who use PCP have been known to engage in unpredictable and sometimes violent acts.

The criminal justice system is also impacted by dissociative drugs, as law enforcement resources are often used to address incidents involving individuals under the influence of these substances. In addition, the cost of healthcare, rehabilitation, and legal proceedings associated with dissociative drug use can be substantial.

In conclusion, we cannot ignore the impact of dissociative drugs on crime and violence. Therefore, it is important for individuals, communities, and the government to be aware of the dangers of these drugs and take steps to prevent and reduce their use. This can include education, treatment, rehabilitation programs, and effective law enforcement measures to prevent the sale and distribution of these dangerous substances.

The Impact of Dissociative Drugs on the Economy

The impact of dissociative drugs on the economy is a significant concern for society. These drugs can lead to various negative consequences, including decreased productivity, increased healthcare costs, and decreased economic growth.

One of the key ways that dissociative drugs impact the economy is through decreased productivity. Individuals who use these drugs are more likely to miss work, struggle with concentration and motivation, and have **reduced job performance**. This can result in decreased productivity, lower earnings, and reduced economic growth.

Another major impact of dissociative drugs on the economy is increased healthcare costs. These drugs can lead to serious health problems, including respiratory failure, coma, and even death. This can significantly strain healthcare systems and increase government and healthcare providers' costs.

In addition, the cost of law enforcement and criminal justice resources used to address incidents involving individuals under the influence of dissociative drugs can also be substantial. This includes the cost of arrests, trials, and rehabilitation programs.

In conclusion, the impact of dissociative drugs on the economy is significant and far-reaching. Therefore, it is important for governments, communities, and individuals to be aware of these impacts and to take steps to prevent and reduce the use of these dangerous substances. This can include education, treatment, rehabilitation programs, and effective law enforcement measures to prevent the sale and distribution of dissociative drugs.

The Impact of Dissociative Drugs on Healthcare

The impact of dissociative drugs on healthcare is a major concern for society. These drugs can cause serious health problems, including respiratory failure, coma, and even death, putting a strain on healthcare systems and increasing costs for governments and healthcare providers.

One of the key ways that dissociative drugs impact healthcare is through their effects on the central nervous system. These drugs can cause hallucinations, confusion, and impaired motor function, leading to an increased risk of accidents and injuries. These accidents can sometimes result in serious health problems, including head injuries, broken bones, and organ damage.

Another way that dissociative drugs impact healthcare is through their potential to cause long-term health problems. These drugs can result in chronic health conditions, including memory problems, mood disorders, and psychological trauma. This can result in increased costs for healthcare providers and decreased quality of life for individuals who use these drugs.

In addition, the cost of treatment and rehabilitation programs for individuals affected by dissociative drugs can also be substantial. This includes the cost of hospitalization, therapy, and rehabilitation services.

In conclusion, the impact of dissociative drugs on healthcare is significant and far-reaching. Therefore, it is important for governments, communities, and individuals to be aware of these impacts and to take steps to prevent and reduce the use of these dangerous substances. This can include education, treatment, rehabilitation programs, and effective law enforcement measures to prevent the sale and distribution of dissociative drugs.

The Impact of Dissociative Drugs on Education

The impact of dissociative drugs on education is a major concern for society. These drugs can seriously affect individuals' mental and physical health, leading to decreased educational attainment and reduced long-term economic opportunities.

One of the key ways that dissociative drugs impact education is through their effects on cognitive function. These drugs can cause memory problems, confusion, and reduced motivation, making it difficult for individuals to concentrate and perform well in school. This can result in lower grades, decreased educational attainment, and reduced opportunities for future employment and financial stability.

Another way that dissociative drugs impact education is through their potential to cause long-term health problems. These drugs can result in chronic health conditions, including memory problems, mood disorders, and psychological trauma. This can make it difficult for individuals to maintain focus and perform well in school, leading to decreased educational attainment and reduced opportunities for future success.

In addition, the cost of treatment and rehabilitation programs for individuals affected by dissociative drugs can also be substantial. This includes the cost of therapy, rehabilitation services, and missed school days.

In conclusion, the impact of dissociative drugs on education is significant and far-reaching. Therefore, it is important for governments, communities, and individuals to be aware of these impacts and to take steps to prevent and reduce the use of these dangerous substances. This can include education, treatment, rehabilitation programs, and effective law enforcement measures to prevent the sale and distribution of dissociative drugs.

The Role of Government and Law Enforcement in Addressing the Use of Dissociative Drugs

The role of government and law enforcement in addressing the use of dissociative drugs is critical to reducing the negative impacts of these substances on society. These entities play a vital role in preventing the sale and distribution of dissociative drugs and providing education, treatment, and rehabilitation resources to individuals affected by these substances.

One of the key ways that governments and law enforcement can address the use of dissociative drugs is through effective law enforcement measures. This includes cracking down on illegal drug labs and distributors and intercepting and confiscating illegal drugs before they reach users. These efforts can help to reduce the supply of these drugs and make them less accessible to individuals who might use them.

Another way that governments and law enforcement can address the use of dissociative drugs is through the provision of education, treatment, and rehabilitation resources. This includes developing educational programs to raise awareness about the dangers of these drugs, as well as creating treatment and rehabilitation programs for individuals affected by these substances. These resources can help individuals overcome dissociative drugs' negative effects and regain their health and productivity.

In conclusion, the role of government and law enforcement in addressing the use of dissociative drugs is critical to reducing the negative impacts of these substances on society. In addition, effective measures to prevent the sale and distribution of these drugs, as well as education, treatment, and rehabilitation resources, are essential to reducing the harm caused by dissociative drugs and ensuring a safe and healthy society.

Chapter Summary

- Dissociative drugs are psychoactive substances that produce feelings of detachment and disconnection.
- Dissociative drugs have been linked to increased crime and violence and significantly affect the economy and healthcare.
- The impact of dissociative drugs on crime and violence includes violent and dangerous behavior and the use of law enforcement resources.
- Dissociative drugs impact the economy by decreasing productivity, increasing healthcare costs, and straining the criminal justice system.
- Dissociative drugs impact healthcare by causing serious health problems, increasing the risk of accidents and injuries, leading to chronic health conditions and increased treatment costs.
- It is important for individuals, communities, and the government to be aware of the dangers of dissociative drugs and to take steps to prevent and reduce their use.
- These steps include education, treatment, rehabilitation programs, and effective law enforcement measures.
- The impact of dissociative drugs on society is significant and far-reaching. It requires a comprehensive approach to address the issue.

10

DISSOCIATIVE DRUGS AND ADDICTION

A woman stands in the dark, completely enthralled by the small light of her phone. She can feel the pull of the addiction, yet she fights against it, knowing it can bring her nothing but darkness.

Dissociative drugs, also known as dissociatives, are a class of psychoactive substances that alter perception and consciousness. They have been used for centuries for their ability to create an altered state of mind. However, they also carry a high risk of addiction and negative side effects. Therefore, understanding the nature of dissociative drug addiction is crucial to help those affected by this condition. In this chapter, we will explore the physical and psychological effects of dissociative drug addiction, the development and progression of the disorder, the role of social and environmental factors, and the available treatment options. By delving into the science behind dissociative drug addiction, we aim to increase awareness, understanding, and compassion for those struggling with this complex issue.

Overview of Dissociative Drug Addiction

Dissociative drug addiction is a complex condition that involves the chronic use of dissociative drugs such as ketamine, PCP, and DXM. These drugs produce feelings of detachment from one's self, surroundings, and reality. In addition, they are often used for their mind-altering effects. Dissociative drugs block certain neurotransmitters in the brain, leading to changes in perception, mood, and consciousness.

However, the repeated use of dissociative drugs can lead to physical and psychological dependence and tolerance, where higher doses are needed to achieve the desired effects. This can result in a cycle of substance abuse, leading to addiction and damaging consequences on one's personal, professional, and social life.

Dissociative drug addiction can have severe health implications, including memory loss, anxiety, depression, and physical problems such as bladder and kidney damage. It can also increase the risk of developing mental health disorders and substance use disorders, making it a complex and challenging condition to treat.

In conclusion, dissociative drug addiction is a serious issue that can have severe and long-lasting consequences. Therefore, it is crucial to understand this condition's nature and be aware of the dangers of chronic dissociative drug use. The first step towards recovery is acknowledging that one has a problem and seeking help.

The Physical and Psychological Effects of Dissociative Drug Addiction

Dissociative drug addiction's physical and psychological effects can be severe and long-lasting. Dissociative drugs alter the balance of neurotransmitters in the brain, leading to changes in perception, mood, and consciousness. This can result in both short-term and long-term effects.

Short-term effects of dissociative drug use can include hallucinations, changes in perception of time and space, increased heart rate, increased blood pressure, and impaired motor function. In addition, dissociative drugs can sometimes cause dangerous side effects, such as overdose and potentially fatal respiratory depression.

Long-term effects of dissociative drug addiction can include persistent changes in perception, memory loss, depression, anxiety, and other mental health problems. Chronic dissociative drug use can also result in physical problems, such as bladder and kidney damage and cardiovascular problems.

In addition to the physical effects, dissociative drug addiction can also have severe psychological effects. Substance abuse can lead to a decline in one's personal and professional life, causing financial and relationship problems and feelings of hopelessness and helplessness. The cycle of substance abuse and addiction can be difficult to break, leading to a decline in physical and mental health and overall quality of life.

In conclusion, dissociative drug addiction's physical and psychological effects can be severe and long-lasting. Therefore, it is crucial to seek help at the earliest signs of substance abuse to prevent the negative consequences of addiction. In addition, treatment options such as behavioral therapy, medication, and support groups can help individuals overcome dissociative drug addiction and achieve a healthier, more fulfilling life.

The Development and Progression of Dissociative Drug Addiction

Various individual and environmental factors can influence the development and progression of dissociative drug addiction. Substance abuse often starts as a form of recreational drug use. Still, it can quickly escalate into a dangerous cycle of addiction.

One of the key factors in the development of dissociative drug addiction is tolerance. Over time, the brain becomes accustomed to the presence of the drug, leading to a decrease in its effectiveness. This can increase drug use as individuals seek to achieve the same effects they did when they first started using. This increase in drug use can lead to physical and psychological dependence and, eventually, addiction.

Social and environmental factors can also play a role in the development of dissociative drug addiction. For example, substance abuse can often stem from a desire to escape stressful life circumstances, such as trauma, mental health problems, or relationship problems. Substance abuse can also be influenced by peer pressure and access to drugs.

The progression of dissociative drug addiction can be difficult to predict but is often characterized by a decline in physical and mental health and personal and professional life. In addition, substance abuse can lead to financial problems, relationship problems, and a decline in overall quality of life.

In conclusion, the development and progression of dissociative drug addiction are complex and influenced by a range of individual and environmental factors. Understanding these factors can help individuals recognize the warning signs of addiction and seek help before it is too late. In addition, early intervention can help prevent the negative consequences of addiction and increase the chances of a successful recovery.

The Role of Social and Environmental Factors in Dissociative Drug Addiction

Social and environmental factors play a significant role in the development and progression of dissociative drug addiction. Substance abuse can stem from a desire to escape life stressors, such as trauma, mental health problems, or relationship problems. Peer pressure, drug access, and cultural attitudes toward substance abuse can also contribute to the development of addiction.

Peer pressure can be a major factor in the initiation of drug use. It can influence an individual's decision to continue using drugs. Being surrounded by individuals who use drugs or having friends who are also addicted can increase the likelihood of substance abuse and addiction.

Access to drugs is also a major factor in the development of dissociative drug addiction. The availability of drugs and the methods of obtaining them can greatly influence an individual's decision to use and continue using drugs.

Cultural attitudes towards substance abuse can also play a role in the development of addiction. For example, substance abuse may be normalized or encouraged in some cultures, leading individuals to believe that drug use is acceptable or desirable.

In addition to these factors, environmental stressors can also contribute to the development of dissociative drug addiction. For example, individuals who experience trauma, poverty, or other forms of adversity may be more likely to turn to substance abuse as a means of coping.

In conclusion, social and environmental factors play a significant role in the development and progression of dissociative drug addiction. Understanding these factors is crucial to preventing substance abuse and supporting individuals in their recovery journey. Addressing these factors can also help reduce the stigma associated with substance abuse and improve access to effective treatment and support.

The Treatment Options for Dissociative Drug Addiction

The treatment of dissociative drug addiction can be complex and may involve a combination of different approaches. The most effective treatments for dissociative drug addiction typically involve a combination of behavioral therapy, medication, and support from family and friends.

Behavioral therapy is a critical component of dissociative drug addiction treatment. This therapy can help individuals understand and address the underlying factors contributing to substance abuse. Common behavioral therapies used in treating drug addiction include cognitive-behavioral therapy, contingency management, and motivational interviewing.

Medication can also be an effective treatment option for dissociative drug addiction. Some medications can help alleviate withdrawal symptoms, reduce cravings, and support recovery. For example, naltrexone and acamprosate effectively treat addiction to dissociative drugs.

Inpatient and outpatient treatment programs can also provide a supportive environment for individuals in recovery. These programs can offer a range of services, including individual and group therapy, medication management, and support from peers in recovery.

The support of family and friends is also an important factor in the recovery process. Encouragement and support from loved ones can help individuals stay motivated and on track in their journey toward recovery.

In conclusion, the treatment of dissociative drug addiction can be complex. It may involve behavioral therapy, medication, and support from loved ones. Seeking help as early as possible can greatly increase the chances of a successful recovery and improve overall health and well-being. It is important to find a treatment program tailored to individual needs and to work closely with a healthcare provider to ensure a comprehensive and effective treatment plan.

Chapter Summary

- Dissociative drugs are psychoactive substances that alter perception and consciousness and carry a high risk of addiction.
- Dissociative drug addiction is a complex condition involving chronic use of drugs such as ketamine, PCP, and DXM.
- Dissociative drug addiction can have severe physical and psychological effects, including memory loss, anxiety, depression, bladder and kidney damage, and an increased risk of mental health disorders.

- Various individual and environmental factors, such as tolerance, social and environmental factors, and stress, can influence the development and progression of dissociative drug addiction.
- Short-term effects of dissociative drug use can include hallucinations, changes in perception, and impaired motor function. In contrast, long-term effects can include memory loss, depression, anxiety, and physical problems.
- The progression of dissociative drug addiction is often characterized by a decline in physical and mental health and personal and professional life.
- Social and environmental factors, such as peer pressure, drug access, and stress, play a significant role in the development and progression of dissociative drug addiction.
- Early intervention and treatment options, such as behavioral therapy, medication, and support groups, can help individuals overcome dissociative drug addiction.

AFTERWORD

In conclusion, this book has provided a comprehensive overview of dissociative drugs, including their history, chemical properties, medical uses and abuses, risks and side effects, laws and regulations, and personal experiences. It has also explored dissociative drugs' relationship with mental health, society, and addiction.

Dissociative drugs can have both positive and negative effects on individuals and society, and it is important to understand their impact. However, the risks associated with dissociative drugs can be serious, and it is crucial to take precautions to prevent misuse.

Ongoing research is being conducted better to understand dissociative drugs' effects and potential medical uses. In addition, new laws and regulations are being developed to address the misuse of these substances.

Finally, it is important to remember that individuals should only use dissociative drugs under the supervision of a medical professional. It is also important to educate yourself about the risks and dangers of these substances and seek help if you or someone you know is using them.

ACKNOWLEDGMENTS

Writing this book would not have been possible without the help and support of many individuals. First, I would like to extend my heartfelt gratitude to my family and friends, who have always been there for me, providing unwavering support and encouragement throughout the writing process.

I am deeply grateful to the experts and researchers in the field of psychedelics and hallucinogens who have shared their knowledge and insights with me. Their contributions have been invaluable in helping me to understand the subject matter better.

I would also like to acknowledge the individuals who shared their personal experiences and stories with me. Their candid and honest accounts have added a unique and personal perspective to the book.

I also want to thank my editor, who helped me to refine my work and make it the best it could be.

Finally, I would like to thank all the readers who took the time to read this book. This book will contribute to a better understanding of these fascinating and powerful substances and their potential benefits for mental health and well-being.

ABOUT THE AUTHOR

Terence Wright is a researcher, writer, and expert on psychedelics, focusing on hallucinogens' effects and cultural significance. He has studied the subject for several years and has published several articles on the topic. Terence has a deep passion for understanding how psychedelics have been used throughout history and how they can be used in the future. He is an advocate of the responsible use of psychedelics. He believes that with more research and understanding, these substances can positively impact mental health and well-being. When not writing or researching, he enjoys hiking and spending time with his family.

FROM THE AUTHOR

Dear Reader,

I hope you enjoyed the book! I would love to hear your feedback—I personally read every review. If you have three minutes to help me, can you please post a short review on the platform from which you purchased this book?

Also, I would be honored if you would sign up for my mailing list to keep updated on my upcoming books, exclusive content, special promotions, and events. Plus, as a subscriber, you'll have the opportunity to **download my new books for free** in exchange for an honest review. Your feedback is important to me and will help me improve my books and reach a wider audience. Scan to QR code to sign up.

SCAN ME

Best, Terence Wright

Made in the USA
Monee, IL
04 November 2026

67379454R00066